HANDBOOK OF OREGON
PLANT AND ANIMAL FOSSILS

William N. Orr
University of Oregon
Dept. of Geology

Elizabeth L. Orr

Eugene, Oregon
1981

LC 81-90259
ISBN 0-9606502-0-2

Typeset by Pan Typesetters
Printed in the United States of America
Cover: *Epigaulus*, horned rodent. Illus. by Laura Beckett

Dedicated to the early morning berry pickers.

CONTENTS

MAPS AND CHARTS

MAPS AND CHARTS

PREFACE AND ACKNOWLEDGEMENTS

This book developed from a summer paleontology field course offered for several years at the University of Oregon. Although several good shorter volumes have been written on specific fossil groups, a need has existed for a single non-technical treatment summarizing the spectacular prehistoric life of the State. The coverage here of invertebrate and vertebrate animals as well as of plants implies assistance from other specialists in paleontology/biology.

Very special thanks in this regard are due J. Arnold Shotwell, State of Washington; William D. Tidwell, Brigham Young University; Steven Manchester, Oregon Museum of Science and Industry; and Bruce Mate, Oregon State University. For assistance in obtaining references and critical suggestions we thank E.M. Baldwin, University of Oregon; Bruce Batten, University of California, Berkeley; Gregory A. Miles, Exxon, Los Angeles; and Richard Heinzskill, University of Oregon.

July, 1981

W.N. Orr
E.L. Orr

[Thomas Condon] was the pioneer geologist who . . . caught a first glimpse of Oregon's lands as they rose from the ocean bed. He saw the sea-shells upon her old beaches; watched the development of her grand forests; saw her first strange mammals feeding upon her old lake shores; he listened in imagination to the cannonading of her volcanoes and traced the showers of ashes and great floods of lava. He followed the creation of Oregon step by step all through her long geological history. . . .

E. Condon McCornack, 1928

No state is more richly endowed with the records of earth history. No region in the world shows a more complete sequence of Tertiary land populations, both plant and animal, than the John Day Basin.

R. W. Chaney, 1956

INTRODUCTION

"Paleontology" may be defined as the study of fossils. A fossil in turn is recognized as some trace of prehistoric life. The word "trace" here correctly implies that the fossil need not be the organism itself or even its shell but may only be a mold, imprint, a track or even a burrow. Most fossils are either not at all altered or only slightly altered from the original condition when the animal or plant was living. To be elevated to the status of a fossil, the organism must also have a certain degree of antiquity. Ordinarily this means the animal/plant should have lived prior to the last glacial withdrawal from the Northern Hemisphere about 11,000 years ago. Anthropologists studying prehistoric man accept much younger material as *bona fide* fossils, but we do not propose to treat fossil man in this book. If we consider trace fossils, tracks, trails, burrows, etc., as well as fossil pollen and single-celled plants and animals (algae, protozoa), we find that fossil material is far more common than might be expected. One author has suggested that fossils may be recovered with patience from virtually any sedimentary rock. The occurence of fossils in a given rock is aided by the initial frequency of the organism, its relative "preservability" and the environment in which it is entombed. Optimum conditions for preservation of each group of organisms differ and are dealt with separately in each chapter.

As objects of study, fossils may be looked upon as a type of non-renewable resource for the reconstruction of the history of life and to a lesser degree geologic history. The individual fossil may be treated as an historic artifact much as we regard stone age tools, flint lock fire arms, or antique automobiles. Just as the above objects reflect the creative genius of man, to meet an environmental need, an individual fossil represents the product of trial and error by nature over millions of years of environmental change.

A diversified history of climatic changes, oceanic advances and withdrawals, and tectonic events (continental drift and volcanism) in Oregon has left its imprint on the State in the form of representatives of virtually all groups of fossil plants and animals. The present day Cascades effect a rough split in the younger Tertiary rocks between the non-marine, freshwater accumulations of terrestrial plants and animals, mostly mammals, to the East and marine or oceanic forms, mostly molluscs, to the West. In the eastern part of the State the rainshadow of the Cascades has in-

1

hibited the development of thick mature soils like those of the Willamette Valley and Coast Ranges. The sparce vegetation and lack of thick soil profiles, on the other hand, enhances the accessibility of fossils in the underlying rock. Molluscan fossils of western Oregon seem vaguely familiar to anyone who has gone beach combing. Many of the best fossil sites for molluscs occur in sea cliffs where a direct comparison to modern forms may be made at the site. By studying both the marine and non-marine fossils we are able to reconstruct a composite picture of the environmental setting.

In order to gain a better perspective of Oregon paleontology, it is well to mention several of the early paleontologists who worked in this area. Oregon's best known geologist, Thomas Condon, was largely responsible for stimulating initial interest and developing geological study in the State. He first came to The Dalles, Oregon, in 1861 as a minister for the Congregational Church. Although much has been made of the apparent contradictions between the fossil record and the scripture, Condon as a deeply religious individual and yet keen scientific observer did not perceive any conflict. By 1870 he had made numerous trips from The Dalles to the Crooked River and John Day Valley areas where he collected fossil plant and mammal material. Much of this fossil material was sent to O. C. Marsh of Yale University, J. Leidy in Philadelphia, E. D. Cope of the Philadelphia Academy of Sciences, as well as to interested persons at the Smithsonian Institution in Washington, D. C. In 1872 Condon was appointed the first State Geologist, and in 1876 he moved to Eugene accepting an appointment at the University of Oregon. At this time he began to focus more on the geology of the western part of the State.

Many collectors who had come to Oregon to gather material were personally guided through the collecting areas by Condon. Among them, O. C. Marsh collected eastern Oregon sites; C. H. Sternberg collected at Fossil Lake sending the material to E. D. Cope who visited the site himself in 1879. Cope probably visited the John Day area also at this time. Condon loaned his extensive collection of bird fossils from the Fossil Lake area to Cope and this along with Sternberg's collection was taken to Washington, D. C., for identification. Ten years later Shufeldt of the Smithsonian Institution replied to a query of Condon by saying he was still using the collection for study. The many publications of Shufeldt and Cope were the result, Cope himself publishing close to 40 papers on his work in Oregon. J. C. Merriam of the University of California came to the John Day basin in 1899 as the result of correspondence with Condon. Merriam spent that summer collecting with several resulting papers on individual species. Merriam also collected in western Oregon, and his students, Furlong, Stock, and Chaney, continued to work in the State for many years. In 1890 Condon began a correspondence from Eugene with W. H. Dall and J. S. Diller of the Smithsonian Institution. The result was an expedition to Oregon and California by these men for an initial look at the fossil molluscs in the western parts of both states.

The first book on Oregon geology, *The Two Islands*, was published by Condon in 1902.

OREGON FOSSIL PLANTS

Oregon's plant fossil record stands out in the State for the amount and quality of material it has contributed to our knowledge of Tertiary floras. Although a few late Carboniferous (Pennsylvanian) plant assemblages and scattered remains of Mesozoic plants are known from the State, the real value lies in the Tertiary floras. A wide range of paleoenvironments from alpine to low-lands, xeric (dry) to mesic (wet) are to be found here. Excellent floras of well-preserved, diverse plant assemblages are known from the Eocene, Oligocene, and Miocene but are absent or limited from the Pliocene and Pleistocene sequences in areas both east and west of the Cascade range.

Like mammals and invertebrates, plants present very special problems of preservation in order to get into the fossil record. The former fossil types with bone, teeth and shell of mineral material only need to be buried and escape solution by percolating groundwater to earn a place in the fossil record. Plants, on the other hand, are susceptible to decay, bacteria, oxidization, and insects, depending on whether the plant remains in question are wood, leaves, or pollen. The grain size of the entombing sedimentary rock also may contribute to or preclude fossilization for plants. Coarse-grained rocks such as sand or gravel (conglomerate) represent areas of rapid deposition and would seem to be an ideal environment for fossil plant preservation. On the contrary, delicate leaves fossilize best only if they are pressed flat. This condition is typical of lake sediments with fine-grained clay size particles such as the muds and shales. Coarse-grained sediments such as sand or gravel also permit the passage of water through the rock after burial. These and other fluids may dissolve and oxidize leaves, wood, or pollen, or they may enrich or fossilize plant remains with secondary mineralization. Ordinarily pollen is preserved intact. That is, it undergoes no substantive change as it exists in the fossil record other than being flattened. The waxy, resinous wall of pollen grains are especially suited to preservation, but they may be oxidized unless they are buried rapidly. Leaves are most frequently preserved as a carbon film in the entombing rock. This type of fossilization, called "distillation," is the simple process of removing all the volatile organic material from the leaf without disturbing the basic shape. What remains after distillation is a fine carbon film of the leaf with the outline, veins and even cells preserved. Again, the latter process has added nothing chemically to the fossil but has instead slowly removed portions leaving behind a thin black carbon residue. Fossil wood is preserved in a variety of ways. Initially the wood may be distilled like a leaf by the slow removal of all but the carbon. More often, wood undergoes a treatment called "permineralization." This process entails the infilling of all the vacant pore spaces by mineral material, usually either calcite or some form of silica. Fossil wood in this condition may look unaltered, but it has a distinctly "heavy" feel due to the extra mineral material it contains. The unaltered appearance of wood in this condition has been dramatically displayed when individuals, thinking such wood is unfossilized, have actually attempted to run it through a sawmill, destroying the sawblade in the process. Fossil wood in the permineralized state may take a polish on a lapidary wheel, but often the woody tissue makes the specimen crumbly. When individuals

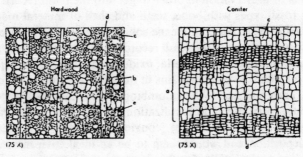

Cross sections of typical conifer and hardwood: a) annual growth; b) vessel, c) fiber; d) multiseriate ray; e) parenchyma.

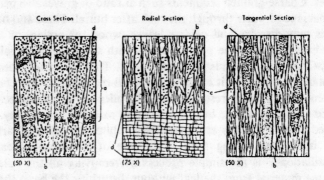

Three standard views of Platanus (sycamore) wood: a) annual growth; b) vessel; c) fiber; d) multiseriate ray.

after Eubanks, 1960

6

refer to "petrification" or to a "petrified" piece of wood they are usually describing wood that has undergone complete replacement by mineral material, usually silica in the form of calcedony, jasper, or agate. This slow process permits the preservation of delicate features of the xylem of the wood including annual rings and details of individual cells. Like permineralized wood, this replaced wood or petrified wood is noticeably heavier than ordinary wood and will usually take a nice polish. The shape of wood and leaves may also be preserved as impressions in the form of casts or molds.

With the difficulties ennumerated above including decay and insect attack, one might expect that the plant fossil record would be dismal. On the contrary, plants are generally better represented than, for example, mammals. The success of plants in the fossil record is one of numbers. Whereas a mammal will produce as potential fossil material around 100 separate bones in its entire skeleton, the typical deciduous or hardwood tree will produce annually tens of thousands of leaves and possibly one hundred times that in its entire lifetime. The amount of pollen produced by one plant on either a lifetime or annual basis can be estimated in the millions. Leaves and wood easily float on water and are carried with a minimum of abrasion and wear to all manner of burial sites prior to entombment. Pollen may be airborne for weeks before it is carried by water to a burial site. We see, then, in the plant fossils a successful numerical strategy of winning a place in the fossil record. The chemistry of plant tissues also contributes to their good fossil record. Pollen, wood, leaves and fruit all contain organic compounds that are resistant to decay.

The plant fossil record is the result of a complicated equation of plant habitats, structure, preservation and physiology. Large thick leaves, for example, are more easily stream carried and preserved than thin, small ones. Most needle trees do not annually shed leaves, but their sum total of leaves exceeds any deciduous type. Dispersal of needles by wind, furthermore, is not nearly as thorough as that for deciduous types.

One of the most useful aspects of plant fossils is their utility as an index to prehistoric climates. The flora that develops in a given area is the delicate by-product of a complex and interrelated series of environmental factors including, in part, rainfall, temperature, sunlight and growing season length, and soils. Often when fossil plants are used to determine the geologic age of a rock or to "correlate" from one area to another, on careful inspection, the geologist may be really correlating climates, not time. This problem can be made to work to the geologists' advantage if a good inventory of the various climatic events that have occurred in a given geographic area is available. If such an inventory or climatic chronology is available, the geologist may then just look for specific climatic events in the fossil record to date the unknown sample. Detling (1968) and Wolfe and Hopkins (1967) have summarized the climatic events of the Pacific Northwest from the Cretaceous to the Holocene.

In floral sequences, climatic changes are profoundly more evident than the slow evolutionary changes of plant floras through time. In addition to climates, plants may provide a clue to prehistoric physiography of a given area. Axelrod (1966) diagrams altitude successions from hardwood in lowlands to mountain conifer floras at higher altitudes to estimate the altitude at which a given fossil flora was developed. Using this technique he has suggested, for example, that the Miocene Sucker Creek flora of southeastern Oregon was developed in a lowland area of around 500' elevation. The Mascall Miocene flora is estimated to have been developed at around 1500', the Blue Mountain flora at 2000', and the Trout Creek flora around 2400'.

Several of the means for distinguishing living tropical and temperate dicotyledonous plants are directly applicable to fossils. Bailey and Sinnott (1916) after examining the margins of plants from different modern climates observed that leaves with entire or smooth margins are overwhelmingly dominant in lowland tropical areas. Conversely, leaves with non-entire margins dominate in mesophytic cold-temperate areas. Within the tropics non-entire margin leaves favor moist uplands, equable or even environments, and protected cool areas. In cool temperate climatic areas entire margins are favored by arid or dry environments. Chaney and Sanborn (1933) used these observations to conclude that the dominance of entire margin leaves in the Goshen flora suggests tropical conditions; whereas, the nonentire margin leaves of the Bridge Creek flora suggest a more temperate environment.

Krasilov (1975) reviewed the comparison of tropical/temperate plant morphology and was able to show distinctions between root structure, leaf size, overall leaf shape, leaf structure (simple vs. compound), venation pattern (pinnate vs. palmate), leaf thickness, leaf length and cell structure. Several paleobotanists have made similar observations on modern plants to deduce paleoecology of fossil floras. Axelrod and Bailey (1969), for example, observed that small-serrate leaves dominate in the cold

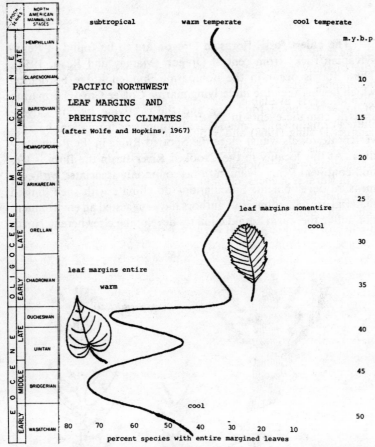

PACIFIC NORTHWEST
LEAF MARGINS AND
PREHISTORIC CLIMATES
(after Wolfe and Hopkins, 1967)

zone of the Northern Hemisphere and large smooth edge leaves dominate in the Southern Hemisphere.

The danger of making paleoclimatic conclusions of this type is the assumption that climate is the only variable affecting leaf morphology or other leaf characteristics. To effectively make such conclusions the reason for the change in plant morphology from the tropical to temperate areas must be clearly understood. Chaney suggests that the difference between entire and non-entire leaves is one of leaf drip points and therefore fundamentally controlled by the available moisture. It should also be pointed out here that some factors such as leaf thickness will affect the fossil record. For example, a thick coarse tropical leaf might preserve well in the fossil record within a variety of rock types. Small very fine thin leaves of temperate climates, on the other hand, would be less likely to be preserved in any but the finest clay muds under very low energy current systems. A flora, then, with mixed temperate (non-entire) and tropical (entire large) leaves might after fossilization appear to be dominantly tropical.

9

The oldest fossil floras in Oregon are to be found in Pennsylvanian rocks from central Oregon (Mamay and Read, 1956). These fossils occur in the nonmarine Spotted Ridge Formation which lies between the underlying marine Coffee Creek Formation of lower Carboniferous Age and the overlying marine Coyote Butte Formation of Permian Age. Both marine formations bear invertebrate fossils which place the Spotted Ridge in the Pennsylvanian. At this locality in the Crooked River Basin the flora is like mid-continent Pennsylvanian floras commonly associated with coal measure development. Speculations on climate with such limited material are hazardous, but authors have suggested an environment not unlike the warm humid climates developing elsewhere in North

PENNSYLVANIAN

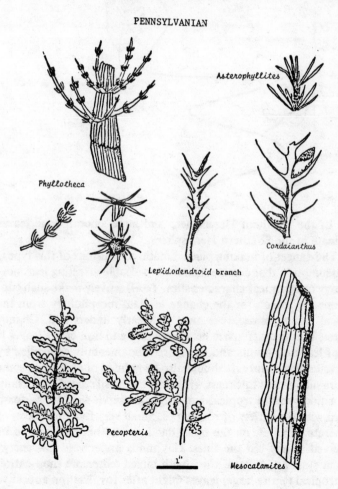

Asterophyllites

Phyllotheca

Cordaianthus

Lepidodendroid branch

Pecopteris

1"

Mesocalamites

America at the same time. Chaney (1956) believes the flora here is of an upland type, rather than lowland, because of the lack of actual coal development. Plant material from this locality is reviewed by Arnold (1953) who was able to add the common fern-like foliage, *Pecopteris*, to an earlier list (Read and Merriam, 1940). In a summary paper (Mamay and Read, 1956) the floral list was systematically revised and described to include the following genera: a lepidodendroid, *Asterophyllites* (ancestral joint grass), *Mesocalamites* (ancestral joint grass), *Phyllotheca* (ancestral joint grass), *Pecopteris* (fern-like), *Cordaianthus* (gymnosperm cone), *Schizopteris* (fern), *Dicranophyllum* (fern-like), and *Stigmaria* (a *Lepidodendron* root-stock).

Oregon floras of fern and fern-like plants in the Pennsylvanian and Jurassic bracket the Permian with its locally adverse climates and periods of extended drought. Triassic floras are unknown in Oregon.

Jurassic floras are described from Douglas County near Riddle in the southwestern area of the State. As with the Pennsylvanian material, fossil invertebrates have been used to corroborate the Jurassic date for these floras. These plant collections from the Riddle Formation at Buck Peak, Thompson Creek and Nichols Station are invariably found in association with *Buchia piochii* (Gabb), an oyster-like clam that occurs in near reef accumulations in Mesozoic upper Continental shelf environments.

The floristic composition in these Jurassic floras is altogether

JURASSIC

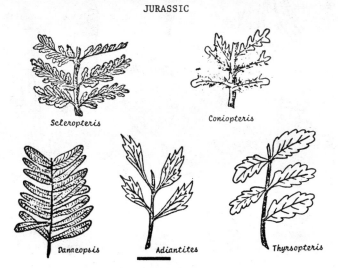

Scleropteris

Coniopteris

Danaeopsis

Adiantites

Thyrsopteris

11

different from that of the older Pennsylvanian floras. Dominant forms include ferns (*Coniopteris, Thyrsopteris, Sagenopteris*) as well as the gymnospermous cycads (*Nilssonia, Ptilozamites*), conifers

JURASSIC

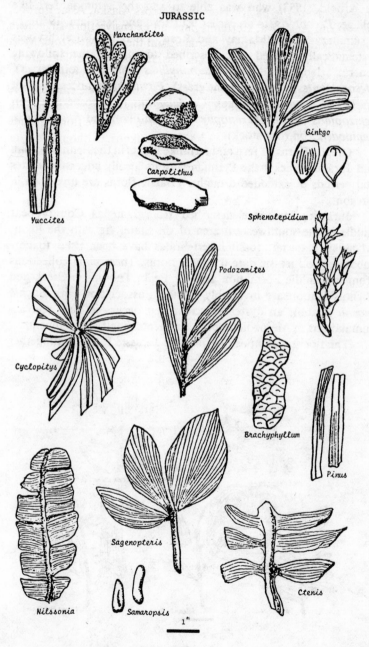

1"

(*Sphenolepidium*, *Pinus*, *Araucarites*) and ginkgos described by Fontaine (1905) and Knowlton (1910). These floras are part of a larger widespread series of mid-Mesozoic floras extending continuously from the Pacific Northwest across Alaska, Asia, Europe, Africa and Australia. Certainly such an extensive flora suggests a wide ranging stable tropical climate from the mid-Mesozoic. Most of the conifers and ginkgos from this Jurassic assemblage are of types now extinct. Cycads occur today in tropical climates only.

Cretaceous floras are rare in Oregon, but some elements have been described from lower Cretaceous units exposed on the Elk River in the Port Orford area (Lowther, 1967). These floras include some six species of ferns, cycads and ginkgos. Although not as cosmopolitan, they bear a high degree of resemblance to the Jurassic floras and probably reflect similar warm moist climatic conditions. Gregory (1969) lists a Cretaceous palm locality in the Mitchell area.

In eastern Oregon, species of the tree fern, *Tempskya*, are found in the Lightning Creek area (Read and Brown, 1937). Known only from the lower Cretaceous, fossils of this curious plant are primarily trunk-like structures with distinctive scattered circular irregular bodies in cross-section. The association of land plants such as *Tempskya* with nearby Cretaceous ammonites has prompted authors to suggest the prehistoric presence of an island in the center of a large Cretaceous inland marine sea extending from British Columbia, into northern California, and across eastern Oregon. The incorporation of Continental Drift Theory into this paleogeographic picture may significantly complicate what appears to be a relatively simple setting.

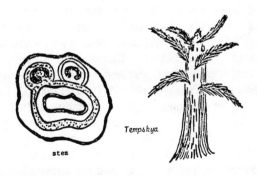

stem   *Tempskya*

No Paleocene plants are described in Oregon, but excellent Eocene floras are known from several localities. Best known of the Eocene floras is from the Clarno Formation. Elements of this flora reflect the development of a widespead forest in the humid subtropical climate of eastern Oregon during late Eocene times. The varieties here of large thick-leafed plants from subtropical zones including *Magnolia* (magnolia), *Cinnamomum* (camphorwood), *Persea* (avocado), *Nectandra* (laurel family), *Meliosma* (sabia family) and *Ficus* (fig) support this. Clarno flora conifers are limited in number but include *Pinus* (pine) and *Glyptostrobus* (water pine). Members of the Lauraceae are the most numerous family in the Clarno flora.

Well-known to paleobotanists as well as to amateur collectors are the so-called "Nut Beds" of the Clarno Formation east of Clarno, Oregon. McKee (?1970) has identified some six genera of fossil fruits and seeds from the tuffaceous clays and sands exposed in the quarry and regards the flora as reflective of a tropical environment. Well-preserved fruits and seeds are rare elsewhere in the fossil record, and one of the better know localities for fruits and seeds is the Eocene London Clay of Britain. Specimens of nuts and fruits are well illustrated by Peterson (1964), Bones (1979), and Scott (1954). The latter author draws interesting parallels between sedimentation processes operating at the Clarno site and modern sedimentation at Paracutin, Mexico, as the result of the volcanic terrain there. Evidence that Scott presents suggests the Clarno beds are water lain tuffs, not the product of ash falls. Both the Clarno and Paracutin areas show evidence of very rapid deposition of ash derived from the erosion of nearby ash beds. Scott concludes that the plants producing the nuts and fruits grew by a streambed, and the latter may have fallen directly into the stream only to be deposited downstream at a spot where the current slowed. Clarno nuts and fruits are dominated by *Juglans* (walnut), the oldest specimens of this genus yet described. As with the London Clay flora, about two-thirds of the Clarno flora is extinct. Genera include three tropical vines, *Parthenocissus*, *Odontocaryoidea*, and *Chandlera*, as well as *Laurocarpum* (Laurel family). Lack of mammal fossils here has made precise correlation difficult, but Stirton (1944) described a species of *Hyrachyus* (rhinoceros) from the Formation as Eocene and later refined this to "probable middle Eocene." McKee regards the lower two-thirds of the Clarno Formation as middle/upper Eocene and the upper Clarno Formation with the Nut Beds and Mammal Quarry as lower Oligocene. Some

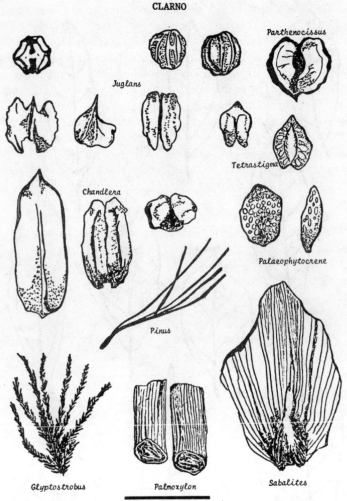

Parthenocissus

Juglans

Tetrastigma

Chandlera

Palaeophytocrene

Pinus

Glyptostrobus   Palmoxylon   Sabalites

evidence also exists that brontotheres recovered from the Mammal
Quarry are species restricted elsewhere to the Oligocene.

The Clarno Formation, in addition to yielding nuts and seeds,
and a prolific subtropical leaf flora, commonly bears fossil palm
leaves, *Sabalites*. A summary paper by Gregory (1969) on fossil
palm leaves (*Sabalites*) and palm wood (*Palmoxylon*) in Oregon
records nine occurrences in the Eocene, one in the Cretaceous, and
two in the Oligocene. The use here of "form genera" *Sabalites* and
Palmoxylon for palm leaves and wood, respectively, is a common
taxonomic strategy in paleobotany where the various plant parts
become separated in the fossil record and cannot be associated with

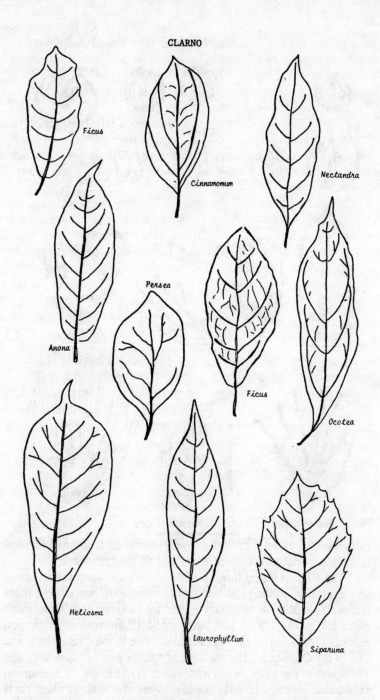

CLARNO

Ficus

Cinnamomum

Nectandra

Persea

Anona

Ficus

Ocotea

Heliosma

Laurophyllum

Siparuna

each other. Eocene localities for palm are almost exclusively in northeast Oregon and dominantly contain wood. Palm species were common in the Eocene but were limited to the warm coastal climates by Oligocene time. Post Oligocene occurrences of palm in Oregon are limited to the Eagle Creek flora.

Arnold (1964) has reviewed the numerous fern families in the Clarno Formation. He notes that because they preserve well and frequently have the reproductive parts attached to the vegetative parts, ferns are often better known than the deciduous plants where the seeds separate from the parent plant.

Elsewhere in western Oregon an Eocene flora collected by Diller (1899) near Coos Bay at Riverton in the upper Coaledo Formation was identified as late Eocene. Hopkins' (1967) analysis of pollen from the Coaledo Formation reveals an Eocene climate warmer and more humid than today in the Coos Bay area reflected by such species as *Magnolia* (magnolia), *Ficus* (fig), *Laurus* (laurel), *Juglans* (walnut), *Rhamnus* (buckthorn), and *Sabalites* (palm). This subtropical climate with 50-60 inches of rainfall annually was characterized by highlands surrounding a basin of deposition, but topography was not as rugged as today. Dott (1966) hypothesizes an open embayment for the Coos Bay area with a low swampy coastal plain adjacent to forested uplands.

The Comstock flora from the lower portion of the Fisher Formation of west central Oregon was described by Sanborn (1937). This late Eocene flora reflects a warm, moist, tropical climate found in association with marine late Eocene shallow water molluscs. The most common plant genera are the tropical *Cinnamomum* (camphorwood) which comprises nearly one-fourth of the total specimens, *Aralia* (aralia), *Magnolia* (magnolia), *Astronium* (cashew family), *Lonchocarpus* (legume family), and *Allophylus* (soapberry family). Wolfe (Peck et al., 1964) notes the occurrence of plant material in the Eocene Colestin Formation where it is exposed on the western flanks of the Cascade Range, but he does not present floral lists.

Middle and late Eocene plant fossils are briefly mentioned from the northern Willamette Valley, Nehalem River basin . (Warren and Norbisrath, 1964). Leaves of the genus *Aralia* (aralia), *Equisetum* (horsetail), *Nymphoides* (floating heart) and ferns were found in association with marine invertebrates and microfossils in probable Eocene Cowlitz Formation. Large tropical leaves of this type of *Aralia* (aralia) are selectively preserved but probably reflect the true environment here. Plant remains associated with molluscs

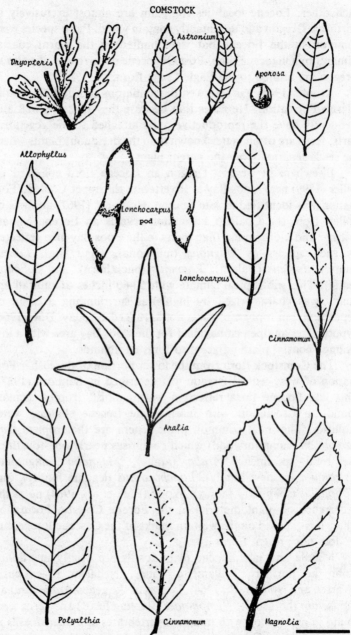

in exposures of the late Eocene Cowlitz Formation west of Timber
were mentioned by the above authors, but no description was
given.

Baldwin (1964) reported a small tropical flora in association with the Rickreall Limestone Member of the Eocene Yamhill Formation. Exposures in the Oregon Portland Cement Company quarry yield species of *Cinnamomum* (camphorwood) and Polypodiaceae (fern).

Oligocene floras are as well represented in Oregon as those of the Eocene. The subtropical conditions of the Eocene continue on into the early Oligocene. Later in the Oligocene more temperate climates prevail as a cooling and drying trend is evident in the plant floras of late Oligocene age. Plant species appearing in the late Oligocene undoubtedly migrated downslope from higher altitudes as the tropical floras disappeared from the lowlands. Although many narrow leaf forms were present, the Oligocene forests were essentially a broadleaf flora including the common deciduous types: *Acer* (maple), *Alnus* (alder), *Betula* (birch), *Carya* (hickory), *Castanea* (chestnut), *Fagus* (beech), *Liquidambar* (sweetgum), *Populus* (cottonwood), *Quercus* (oak), *Salix* (willow), *Tilia* (basswood), and *Ulmus* (elm).

Brown (1959) produced a composite list of Cascade, upper (late) Oligocene plants from various localities in Lane County, the Willamette Valley, Bridge Creek and the Crooked River area in eastern Oregon. He points out that an overwhelming proportion of Oligocene plants were adapted to a moderate temperate climate with adequate precipitation to support a mesophytic forest cover or one reflecting dryer conditions. This trend shown here from tropical conditions in the Eocene and lower Oligocene to temperate conditions in the late Oligocene continues into the Miocene where conditions become even cooler.

The Sweet Home flora correlative with the Eugene Formation described by Richardson (1950) represents one of the earliest Oligocene leaf floras in the State. Leaves preserved in tuffs interfingering with the Eugene Formation include *Sequoia* (redwood) and *Taxodium* (bald cypress) as well as several deciduous hardwoods characteristic of stream-bank, bottomland environments. Petrified wood from the Sweet Home locality bears pseudomorphs of quartz after halite. This occurrence of quartz in the crystal shape of the mineral halite represents the replacement of the salt (halite) crystals by the more stable mineral quartz. Staples (1950) has interpreted the original occurrence here of halite as an indication of a highly saline environment which may have developed when an arm or inlet of the Oligocene ocean was restricted or cut off from the open sea. Evaporation of this isolated water body proceeded to the

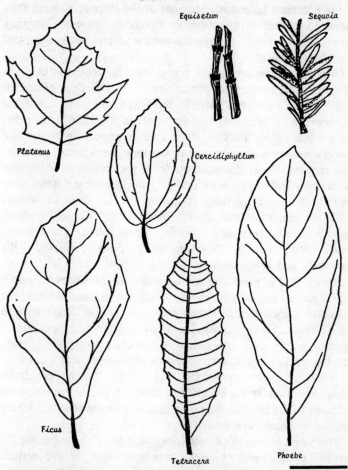

Equisetum

Sequoia

Platanus

Cercidiphyllum

Ficus

Tetracera

Phoebe

crystallization of halite crystals in the water-soaked wood. Wolfe (*in* Peck et al., 1964) records two fossil plant localities near Sweet Home and designates them as from the early Oligocene, Little Butte Volcanics. Gregory (1968) records over 50 species of fossil wood from a collecting site between Sweet Home and Holley. She notes that many of the species are common in the Eocene and are distinctively tropical in nature: *Cinnamomum* (camphorwood), *Schima* (tea family), *Magnolia* (magnolia), and *Ocotea* (lancewood). Most of the trees were preserved *in situ* where they were covered by volcanic ash, but an Asiatic looking portion of the flora may have been transported into the area by the adjacent ocean as drift logs.

Four genera of plants from the Keasey Formation at Mist, Oregon, are listed by Moore (1976) including *Quercus* (oak), *Myrica* (myrtle), *Thuja* (arborvita), and *Ocotea* (lancewood). These occur in association with molluscs and are roughly age equivalent to the Sweet Home flora.

BRIDGE CREEK

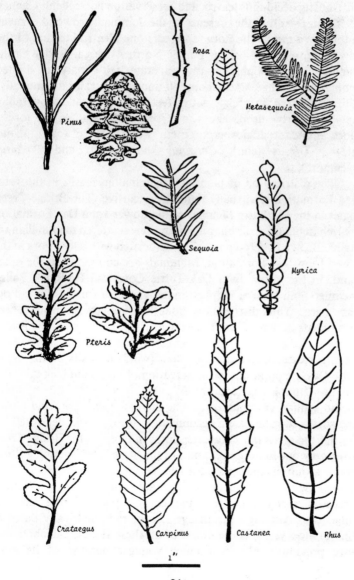

1"

21

Best known of the late Oligocene floras is the Bridge Creek flora of the lower John Day Formation. The Bridge Creek flora and similar Oligocene floras are part of a large inland Oligocene forest suggested by the lack of tropical species. The Bridge Creek is esssentially a *Metasequoia* (dawn redwood) dominated flora of temperate aspect including *Sequoia* (redwood), *Metasequoia* (dawn redwood), and *Taxodium* (bald cypress). Species diversity is high with mostly deciduous leaves, and preservation is excellent. Chaney (1956) suggests that the presence of the dominant redwood element in this flora means the flora represents one identical to that of the now living redwood belt of Pacific North America. Bridge Creek climate was probably warm-temperate with only a limited temperature range and about 40 inches of rainfall annually. Altitude was within a few hundred feet of sea level, and topography included narrow deep valleys to afford protection for the forest. Most abundant angiosperm genera were *Quercus* (oak), *Alnus* (alder), *Ulmus* (elm), *Carpinus* (hornbeam), and *Platanus* (sycamore).

The distribution of plant and mammalian fossils in the John Day Formation is virtually mutually exclusive. The Bridge Creek shales in the Big Basin Member of the Lower John Day Formation are rich in plant fossils but only bear a few scattered mammalian remains. Brown (1959) reports a bat associated with a leaf flora in the lower John Day Formation. Mammals are common, on the other hand, in the upper John Day Turtle Cove and Haystack Valley Members, but little or no plant material except for fossil wood occurs here. This distribution doubtless reflects prehistoric environments of deposition. Lake (lacustrine) sediments of the Bridge Creek shales are conducive to leaf entrapment and preservation but tend to scatter vertebrates. Aeolian (wind) deposits of the upper John Day are optimum for preservation of bones but lack thin, flat laminae ("book page deposition") necessary for leaf preservation. In the aeolian environment, then, the dry leaves are wind blown and scattered rather than accumulated, trapped and deposited.

Of the seven late Oligocene floras in Oregon, the Bridge Creek is the only one in the eastern part of the State. The remaining floras in this group, including the Rujada, Goshen, Willamette Junction, Lyons, Scio and Thomas Creek, are situated in the Western Cascades along what was the coastal plain during the late Oligocene. Although the Bridge Creek, radiometrically dated at 32.5 million years, is the oldest of all these floras, paleobotanists have previously placed it much younger because of its more

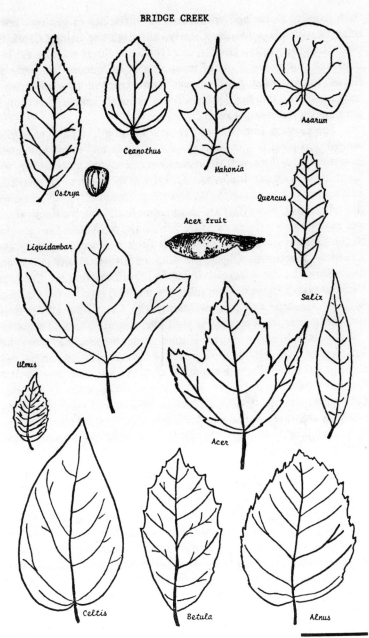

Ostrya
Ceanothus
Mahonia
Asarum
Quercus
Liquidambar
Acer fruit
Salix
Ulmus
Acer
Celtis
Betula
Alnus

temperate floral characteristics. The assumption here that Pacific
Northwest tropical environments invariably give way to temperate
climates automatically places temperate floras above or younger

than tropical floras and epitomizes the difficulty of geologic correlation with plant fossils. Clearly, although the Bridge Creek is older than the western Oregon late Oligocene floras it is also far inland well out of reach of the mild oceanic influence. The western floras on the other hand appear more tropical by comparison because of the increased rainfall and reduced temperature range wrought by oceanic proximity.

The Goshen flora of west central Oregon is typical of a low latitude, tropical to subtropical rain forest in an inland protected situation. The flora was deposited near sea level as is shown by association with marine molluscs. Of the 49 species represented, *Meliosma* (sabia family), *Allophylus* (soapberry), *Nectandra* (laurel family), *Ficus* (fig) are most common. *Tetracera* (liana vine) and *Aristolochina* (liana vine) are among the most abundant genera in the Goshen flora. The age of this flora has been variously assigned as Eocene to Oligocene. Chaney and Sanborn (1933) in describing the flora regard it as Fisher Formation in the late Eocene/early Oligocene interval. Brown (1959) regards the Goshen, as late Oligocene and suggests along with Lakhanpal (1958) that this flora may be correlative with the Rujada flora. The term "Willamette Junction Flora" is used to denote material from the junction of U. S. Highway 99 north of Goshen, distinguishing this locality from the others in the same area. An absolute date for the tuffs of Goshen is placed at 31.0 m.y.b.p. which is late Oligocene. One interpretation (Baldwin, personal commun.) of these two floras is that they are stratigraphically high in the Fisher Formation, the upper portion of which is the nonmarine equivalent of the Eugene Formation.

An important late Oligocene flora in Oregon is the Rujada flora northeast of Cottage Grove at Lookout Point Reservoir. Baldwin (1976) notes the name "Rujada" is a combination of the names of two loggers, R. Upton and J. Anderson, plus USDA. Most of the elements of this flora are warm, temperate, but a few tropical types and the sparse occurrence of *Metasequoia* (dawn redwood) indicate a coastal influence. Lakhanpal (1958) suggests that precipitation at the Rujada area in late Oligocene times was well distributed throughout the year with 50-60 inches per year. Winter temperatures were 30-40 °F. and summer 70-80 °F. Topography was probably irregular and well-drained being close enough to the ocean and sea level to be affected by its moderating effects. Most of the commonly represented plants in this flora have modern temperate equivalents. Particularly numerous were *Alnus* (alder),

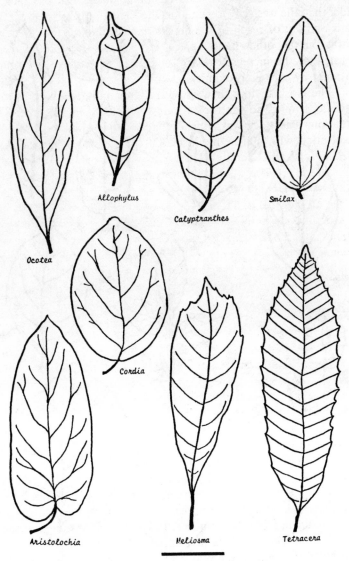

Halesia (snowdrop tree), *Rhus* (sumac), *Quercus* (oak), *Exbucklandia* (witch-hazel family), *Sequoia* (redwood), *Platanus* (sycamore), and the abundant conifer, *Cunninghamia*. Peck (et al., 1964) lists leaf localities in this area as occurring in the Little Butte Volcanic series. The late Oligocene Lyons flora from south of Salem has

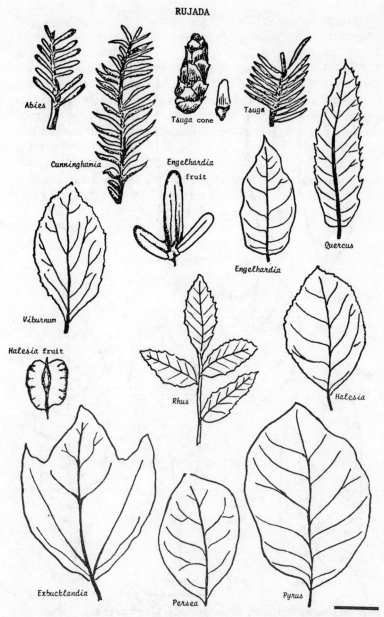

been compared to subtropical and temperate floras on the basis of leaf length, drip point, margin geometry and texture (Meyer, 1973). Lyons flora leaves are dominantly thin, long leaf types over 4″, with non-entire margins that lack drip points. The leaves are lobed

26

or serrate indicating the flora contains both subtropical and temperate species although the temperate species are more abundant. Temperate deciduous Lyons species include *Rosa* (rose), *Metasequoia* (dawn redwood), *Pterocarya* (wingnut), and *Sequoia* (redwood). The presence of subtropical *Cunninghamia* (conifer) and *Meliosma* (sabia family) indicates the forest represented was probably growing near a coastal environment with mild temperatures. The Lyons flora is considered to occur in the Little Butte Volcanic series.

Although close geographically, the Oligocene Scio flora only bears a few elements in common with the Lyons flora. The Scio flora from the "Scio Beds" is dominated by leaves of the abundant *Prunus* (cherry), followed by *Amelanchier* (serviceberry), *Equisetum* (horsetail), *Vaccinium* (blueberry), *Platanus* (sycamore), and *Sequoia* (redwood); and Sanborn (1947) has characterized the climate as warm-temperate to tropical. Coastal proximity of the Scio is suggested by the humid environment, by the presence of tropical species, and by the lack of such deciduous inland types as *Quercus* (oak) and *Acer* (maple) which occur commonly in the Bridge Creek flora.

LYONS

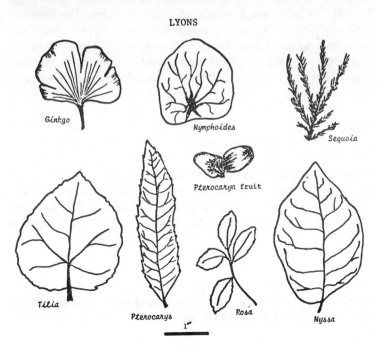

Ginkgo

Nymphoides

Sequoia

Pterocarya fruit

Tilia

Pterocarya

Rosa

Nyssa

1"

27

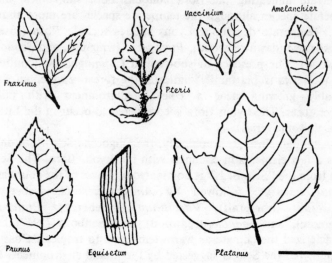

An Oligocene flora from along Thomas Creek and Bilyeu Creek southeast of Salem has been noted by several authors. In the same area as the Scio and Lyons floras this flora includes a large variety of over 20 families and genera of deciduous hardwoods. Tropical to subtropical plants dominate including the common *Siparuna* (siparuna), *Amyris* (rue family), which is the first fossil occurrence of this genus, *Dennstaedtia* (fern), *Metasequoia* (dawn redwood), *Cercidiphyllum* (katsura tree), *Magnolia* (magnolia), *Fissistigma* (tropical vine), *Phoebe* (laurel family), *Tripterygium* (staff-tree family), the first occurrence of this genus in North America, and *Allophylus* (soapberry). The high percentage of lianas and tropical hardwoods in this flora makes it similar to the Goshen, Rujada and other coastal floras of the Oligocene. Annual moisture is estimated at 45-65 inches (Klucking, 1964) with a temperature range from 60 °F in January to 75 °F in July. Peck (et al., 1964) designates the Thomas Creek/Bilyeu Creek flora as mid-Oligocene and puts it in the Little Butte Volcanics, while Klucking (1964) assigns an upper Oligocene age to the same flora and puts it in the Mehama Volcanics, the lowermost member of the Little Butte Volcanics. Eubanks (1960) in an analysis of fossil woods from the Thomas Creek vicinity notes that preserved stumps and logs in groups of twos and threes were probably buried in place. Wood representing some 13 genera is reported of which 3 genera are conifers and the remainder hardwoods.

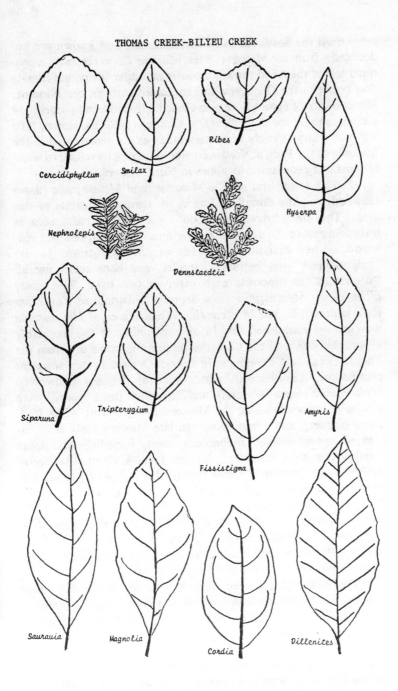

Of all the fossil floras in the State, the best known are undoubtedly from the Miocene series. Miocene floras represent a continuation of the trend toward a cooler and drier temperate climate that began in the Oligocene and continued through the Pliocene. Floras of the Oregon Miocene are essentially modern in aspect but are remarkable for their diversity. Chaney (1956) has noted, "To see such a large variety of species today as are represented in the Miocene of the Pacific Northwest would require travelling to widely separated geographic localities in North America."

During lower and middle Miocene time *Metasequoia* (dawn redwood) was the dominant tree in the temperate forests of this area. The title "dawn redwood" used for *Metasequoia* is misleading since the genus is neither ancestral to cypress nor redwood. The distinct feature of *Metasequoia* is its "oppositeness"—its branches, needles, and cone scales are all "distichous" or opposite each other in two rows. This easily distinguishes *Metasequoia* from *Sequoia* (redwood) or *Taxodium* (bald cypress). By late Miocene time the incidence of *Metasequoia* was greatly diminished due in all probability to the increasing geographic effect of the Cascades intercepting moisture from the Pacific Ocean and developing in eastern Oregon a more apparent rainshadow. Taxodiaceous plants such as *Sequoia* (redwood), *Metasequoia* (dawn redwood), and *Taxodium* (bald cypress) were gradually replaced during the Miocene by species of *Abies* (fir), *Picea* (spruce), and *Pinus* (pine). In late Miocene there is an increased appearance of herbaceous plants including Malvaceae (mallow family), Umbelliferae (parsley family), Compositae (composites), and Gramineae (grass family) due apparently to severity of the winters over this interval. Miocene floras from several prehistoric climatic environments as well as a variety of altitudes are represented in the State. Many of the sites bearing these floras have the additional stratigraphic control of vertebrate fossils. Miocene floras are found in two separate areas of the State, in east central Oregon and in the Willamette Valley.

The many volcanoes building the Cascade Range distributed volcanic sediments through streams and rivers draining areas in which the Miocene forests were developing. Volcanic material obstructed river valleys creating numerous swamps and lakes aided by heavy summer rains blown in from the Pacific Ocean. The heavily forested slopes contributed prodigious amounts of leaves, pollen and wood which accumulated in these many small water bodies.

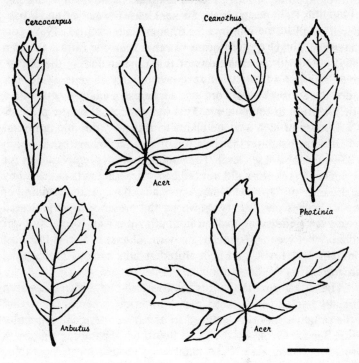

Tuffaceous leaf-bearing sediments referred to as the Alvord Creek Formation in the Steens Mountains are dated (Evernden, et al., 1964) at 21.3 m.y.b.p. Axelrod (1944) interprets the deposition of the Alvord Creek flora as being in a small lake surrounded by moderate topography with distant volcanic cones. Annual rainfall in the immediate vicinity of the lake is estimated at 20-23 inches per year most of which occurred as winter rains. Seasonal temperatures were more moderate than today, and the average yearly temperature was probably 10° higher than today. Members of the Rosaceae (rose family), the most abundant being *Amelanchier* (serviceberry), and *Acer* (maple) occupied the lake shore while *Ceanothus* (buckbrush), *Cercocarpus* (mountain mahogany) and *Juniperus* (juniper) were on the slopes; *Pinus* (pine), *Abies* (fir), and *Pseudotsuga* (Douglas fir) lived a distance from the lake. Remarkably *Quercus* (oak) is not known from the Alvord Creek flora.

31

The Miocene floral series includes the famous Sucker Creek flora found along Succor Creek in Malheur County. First collected in 1900, this flora occurs in the Sucker Creek Formation and is considered middle Miocene in age. Chaney and Axelrod (1959) correlated the Sucker Creek flora with other Miocene floras; Graham (1965) published on all aspects of this flora including the pollen, leaves, fruit, and flowers listing approximately 45 genera. He concluded the Sucker Creek flora was a lake basin environment with a wide variation in topography. This was supported by the presence of a mixture of high altitude plants with those from mid latitudes, and Graham estimates the elevation of the lake to have been as high as 2,000 feet above sea level. The minimum winter temperatures for the Sucker Creek flora did not fall below freezing as is indicated by the presence of *Cedrela* (South American cedar). Annual rainfall of over 43 inches per year is shown by the presence of large leafed *Oreopanax* (modern equivalent in South American forest) and was probably between 50-60 inches per year. *Quercus* (oak) is the most dominant leaf fossil, and high altitude plants such as *Picea* (pine) and *Abies* (fir) are common in the pollen record.

The rock formation in this southeastern Oregon area has been officially named the Sucker Creek Formation, whereas the stream in the same vicinity was changed to its earliest designated popular name, Succor Creek, by the U. S. Board of Geographic Names. A formation retains its official name even though local geographic places may change, hence the two different spellings.

The Rockville flora reported by Smith (1932) occurs in the Sucker Creek Formation exposed near the post office in Rockville, Oregon, and below Sheridan Ridge in the same area. Smith reported nearly 40 species from a dominantly angiospermous flora characterized by north temperate genera: *Carya* (hickory), *Betula* (birch), *Berberis* (Oregon grape), *Populus* (cottonwood), and *Quercus* (oak). Swampy conditions are indicated by an abundance of *Equisetum* (horsetail) and *Typha* (cattail) fragments. Smith considered the flora to be early Miocene or late Oligocene because of the dominance of Oligocene species; however, Wolfe (1969) notes that the Rockville and Sucker Creek floras are distinct and separate and regards the former as middle Miocene.

First collected by Condon and described by Lesquereux (1888) and later by Knowlton (1902) the Mascall flora was assessed by Chaney (1925) as a climax forest representing a large area of low relief and uniform climate, with numerous lakes dammed by volcanic sediments. Fossiliferous localities bearing Mascall floral

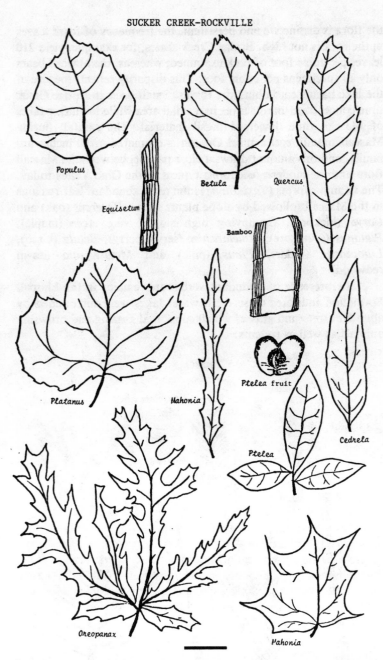

Populus

Equisetum

Betula

Quercus

Bamboo

Platanus

Mahonia

Ptelea fruit

Cedrela

Ptelea

Oreopanax

Mahonia

elements are particularly widespread being found over an area of 700 miles north and south and 250 miles east and west. Although the flora is distinctive and persistent, the frequency of fossil leaves in the rock is not high. Bridge Creek shales, for example, yield 210 leaves per cubic foot of rock examined; whereas, the Mascall bears only 10 specimens per cubic foot. This disparity reflects the size of the lake basins entombing the fossils. Small lakes in Bridge Creek time concentrate many leaves in a small area while the larger lakes of Mascall time dispersed fossil material. The rainfall during Mascall time in east central Oregon is estimated as 30 inches annually, and this amount draws a close parallel between the Mascall flora and the modern oak forest typical of the Ohio Valley today. The swamp cypress (*Taxodium*) is the most abundant leaf remains in the Mascall followed by slope plants such as *Quercus* (oak) and *Carya* (hickory). Occupying high slopes were *Acer* (maple), *Platanus* (sycamore), *Amelanchier* (serviceberry), *Betula* (birch), *Libocedrus* (cedar), *Pinus* (pine) and *Metasequoia* (dawn redwood).

The presence of abundant vertebrate remains in the Mascall Formation indicates these fresh water lakes were surrounded by abundant trees and grasses which supported grazing and browsing animals as well as rodents.

MASCALL

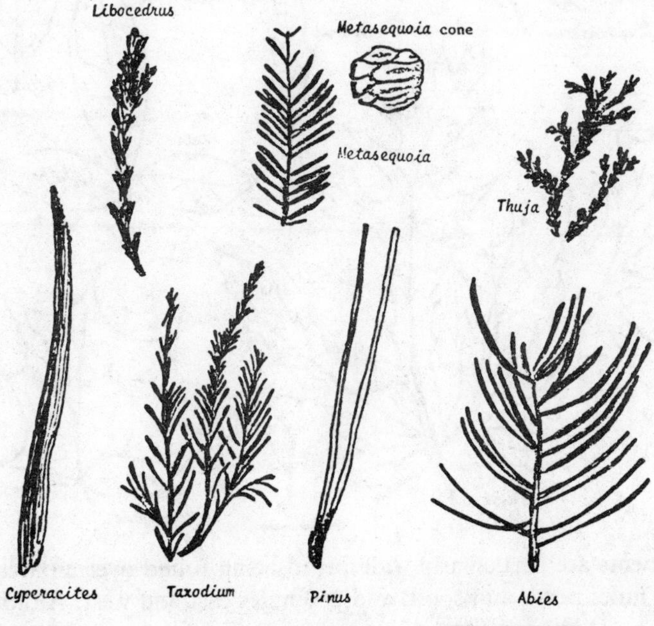

Libocedrus

Metasequoia cone

Metasequoia

Thuja

Cyperacites   Taxodium   Pinus   Abies

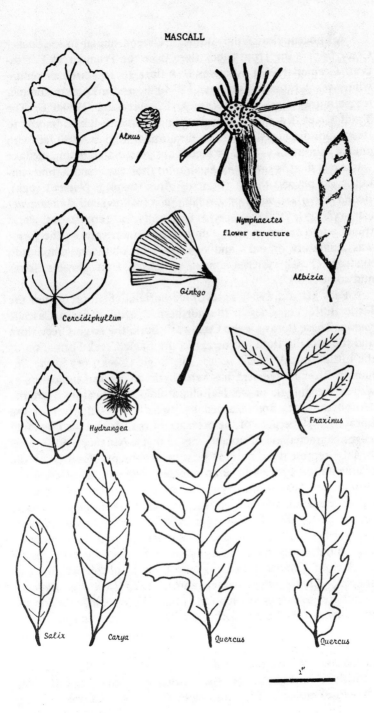

1"

35

A Miocene flora from southeast Oregon similar to the Sucker Creek flora is the Trout Creek flora from the Trout Creek Formation. Graham (1963) estimates this flora to represent minimum winter temperatures of around 35°-40°F, and maximum summer temperatures rarely exceeding 90°F, averaging 75°-80°F. The Trout Creek flora was deposited in an upland lake adjacent to steep, well-drained slopes at an elevation as high as 4,000 feet with annual rainfall of around 50 inches per year. Surrounding the lake was a marsh as is shown by *Equisetum* (horsetail) and *Typha* (cattail), and beyond that a forest of *Acer* (maple), *Quercus* (oak), *Betula* (birch), *Arbutus* (madrone), *Salix* (willow) and *Amelanchier* (serviceberry). Of the 45 species here, only 5 are coniferous, *Pinus* (pine) and *Thuites* (cypress family) being numerous so the forest was a mixture of oaks and conifers with abundant maple and madrone. Fossil material consists of leaves, fruit, flowers, stems and some roots.

Peck (et al., 1964) notes three correlative floras within the Little Butte Volcanics in the northern Oregon Western Cascade region. These floras are the Collawash flora, the Eagle Creek flora and the Molalla flora, occurring in the Eagle Creek Formation of the Little Butte Volcanic series. The Eagle Creek flora in the Columbia River Gorge area has xerophytic or dry leaf indicators as well as mesophytic or wet leaf climatological indicators. Here the xerophytic types are regarded as "true" representatives of the floral ecology because of the presence of maples and the thin large leaves characteristic of upland, dry climates (Chaney, 1918). This locality represents a natural mixture where peculiarities of the sedimentation process have co-mingled elements of different environments. That is, the dry leaf types were brought down from a higher elevation and mixed with those in the valleys. Overall climate during this time was regarded as temperate, somewhat drier and warmer than the present day temperature, and this flora has a large number of species in common with the Oligocene Bridge Creek flora. *Quercus* (oak) is the most abundant followed by *Acer* (maple), *Liquidambar* (sweet gum), *Salix* (willow), *Platanus* (sycamore) and *Populus* (cottonwood). Wolfe (1960) reports the abundance of the genera *Ginkgo* (ginkgo) and *Cephalotaxus* (plum-yew) in the Eagle Creek flora and the remarkable absence of taxodiaceous types. As with the Bridge Creek, the ages various paleobotanists have published for the Eagle Creek flora show the difficulties in making age assignments on climatologically controlled assemblages. The Miocene Eagle Creek flora was first cor-

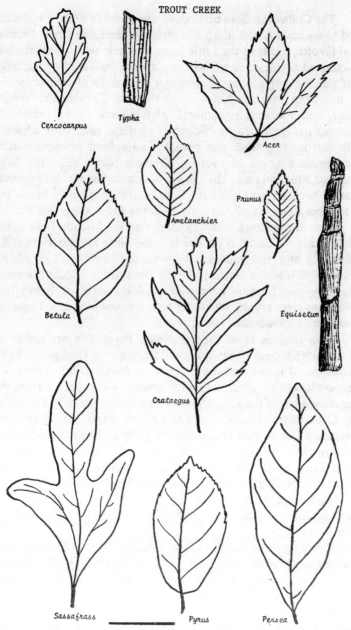

Cercocarpus

Typha

Acer

Amelanchier

Prunus

Betula

Equisetum

Crataegus

Sassafrass

Pyrus

Persea

related with the Clarno flora as Eocene (Chaney, 1918); later it was listed as Oligocene and correlated with the Bridge Creek flora (Chaney, 1920, 1927).

The Collawash flora occurs near the junction of the Collawash and Clackamas Rivers in what is probably the Eagle Creek Formation (Wolfe, 1954) of the Little Butte Volcanic series. This flora is dominated by mesic or wet loving plants as indicated by the large leaf size. Swamp cypress (*Taxodium*) is the most abundant species of this flora which includes *Nyssa* (tupelo), *Liquidambar* (sweet gum), and *Platanus* (sycamore). *Metasequoia* (dawn redwood), *Quercus* (oak) and *Carya* (hickory) also occur suggesting adjacent hills surrounding what was probably an upland swamp environment with a grove of cypress, all these plants requiring large amounts of warm rain. The flora is dominated by warm temperate plants. Because the Collawash flora is stratigraphically below the Columbia River Basalt, an early Miocene age is suggested.

Near Butte Creek and Abiqua Creek, the Molalla flora in the Little Butte Volcanics is divided into the older Molalla flora (12.9 m.y.b.p.) and the younger Weyerhauser flora (12.2 m.y.b.p.; Wolfe, 1969) which would place the flora in the middle Miocene. This lowland flora is composed predominantly of broad leaf deciduous trees among which are Juglandaceae (walnut), Fagaceae (oak) and *Liquidambar* (sweet gum).

Two florules from the John Day Formation are noted by Wolfe (1960). Near Maupin a small florule with *Ginkgo* (ginkgo), *Metasequoia* (dawn redwood), *Quercus* (oak), *Ulmus* (elm), *Liquidambar* (sweet gum), and *Acer* (maple) was collected from the upper member of the John Day and regarded as equivalent in age to the Collawash, Molalla, and Eagle Creek floras. At Monument, Oregon, a florule with *Betula* (birch), *Quercus* (oak), *Sophora* (pea family), *Acer* (maple), *Ulmus* (elm), *Zelkova* (keaki tree), and *Cocculus* (snail seed) was collected from a unit stratigraphically between the John Day and Columbia River Basalts. This latter florule is regarded as younger than the Collawash, Eagle Creek, and Molalla floras and correlates with the Fish Creek florule of the Sardine Formation.

The Blue Mountain flora from three localities in Grant County was reported by Oliver (1934) and later by Chaney and Axelrod (1959) who considered it equivalent to the Mascall. The volcanic tuffs in which the flora occurs interfinger with the Columbia River Basalt of middle Miocene age; however, a formational name is not assigned to this unit. Fossil material consists of leaf and fruit impressions of both trees and shrubs. Seven conifers and 25 deciduous plants are represented with a total assemblage resembling a temperate flora of plants inhabiting the shores and slopes of an

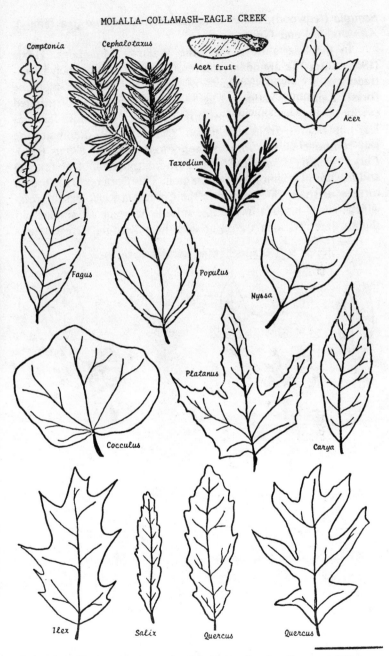

upland lake. The most common forms in the flora are *Quercus* (oak), *Fagus* (beech), *Crataegus* (black hawthorn), *Betula* (birch),

Sequoia (redwood), *Platanus* (sycamore), *Gordonia* (tea family), *Alnus* (alder) and *Pinus* (pine).

In the Sparta flora from the Wallowa Mountains, Hoxie (1965) notes the abundant coniferous pollen from which he extrapolates a paleoenvironment of an upland or slope montane forest. Rainfall is estimated as 50-60 inches per year distributed evenly throughout the year with mild winters. The flora includes leaf impressions, fruits and pollen. *Quercus* (oak) dominates the leaf components, but *Pinus* (pine), *Picea* (spruce), *Ulmus* (elm), *Carya* (hickory), *Salix* (willow), *Populus* (poplar), *Acer* (maple) and *Platanus* (sycamore) were common. This flora occurs in brown tuffaceous shales interbedded in the Columbia River Basalt group which would place it in the late middle Miocene. A similar but slightly later flora in the same area near Keating, Oregon, was

BLUE MTN.-STINKING WATER-SPARTA

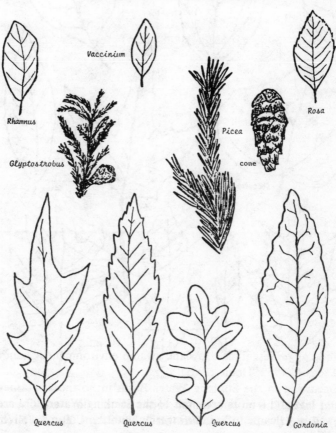

40

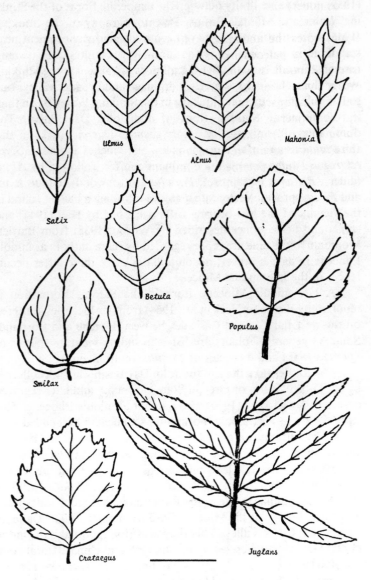

described as representing a lowland swamp environment (Chaney and Axelrod, 1959; Hoxie, 1965).

Occurring in the Stinking Water Basin in northeast Harney County, leaf and fruit impressions of the Stinking Water flora are preserved in volcanic tuffs interfingering with the Columbia River

Basalt, placing the age at middle Miocene. Chaney and Axelrod (1959) note the similarity between the temperate floras of the Stinking Water and Mascall Basins. Paleotopography of the Stinking Water area is estimated to have been a lowland environment near sea level in an area of good drainage. Lignites and swamp assemblages common in the Mascall flora are absent in the Stinking Water flora. This suggests better drainage and broader valleys due to less volcanic sediment blockage in the Stinking Water Basin than in the adjacent Mascall flora of the John Day Basin. The dominance of semixeric indicators such as *Quercus* (oak) in this flora suggests a rainfall amount as low as 10 inches annually. *Glyptostrobus* (water pine) is the dominant conifer, followed by *Alnus* (alder), *Platanus* (sycamore), *Populus* (cottonwood), *Ulmus* (elm) and *Picea* (spruce). Vertebrate material including a badger found in the vicinity of the leaf-bearing tuffs described by Hall (1944) and camel and horse bones described by Wilson (1938) from Bartlett Mountain and Rome areas are regarded by these authors as middle Pliocene in age; however, absolute dates place these latter occurrences in the late middle Miocene.

A late middle Miocene flora in the Sardine Formation is reported by Wolfe (*in* Peck et al., 1964) from six localities scattered on the west flank of the Cascades between Eugene and Portland. Some 33 genera of plants are listed including separate species of *Quercus* (oak) and 4 species of *Populus* (cottonwood).

The Rattlesnake flora in the John Day Basin was first collected by W. D. Wilkinson of Oregon State University and later reported on briefly by Chaney (1956) who dated it as middle Pliocene. This flora occurs in Rattlesnake Formation sediments interbedded with volcanics. Absolute dates of 6.4 m.y.b.p. (Evernden et al., 1964) as well as abundant late Miocene mammals in the unit place it in the late Miocene. The few leaf impressions obtained from the Rattlesnake are of *Ulmus* (elm), *Platanus* (sycamore), and *Salix* (willow).

A flora similar in age to the Rattlesnake, but more widespread and better known, is the Miocene flora from north of The Dalles in Chenoweth Creek Valley. This flora was first reported by Condon (1902) and is characterized by a cool, semiarid climate floral composition with *Quercus* (oak), *Acer* (box elder) as a new species, *Amorpha* (indigo plant), *Ulmus* (elm) and *Cercis* (redbud). This flora is represented by only 12 species from localities in the basal portion of The Dalles Formation, and the lack of many broad-leaf deciduous plant genera suggests reduced summer rainfall. The lack of conifers, live oaks and the small-leafed evergreens is suggestive

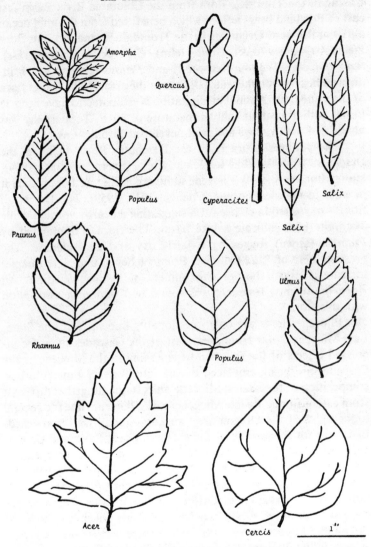

of low altitudes with the plants near stream borders. The Dalles flora shows characteristics of a much more modern flora than that represented by the nearby Eagle Creek flora. Vertebrate fossils of *Aelurodon* (dog), camel remains, and horse teeth (*Hipparion*) (Vanderhof and Gregory, 1940) indicate a late Miocene age for The Dalles Formation.

By comparison with the above late Miocene floras east of the

Cascades, the Troutdale flora from the Columbia River Basin just east of Portland bears a flora which benefitted from the mild ocean air from the west. Occurring in the Troutdale Formation, this flora includes *Quercus* (oak), *Ulmus* (elm), *Cyperacites* (coarse grass), *Salix* (willow), *Sequoia* (redwood) and *Prunus* (cherry) which indicate a heavier summer rainfall than in the area of The Dalles flora of the same time. Annual precipitation is estimated to have been 35 inches with a mean annual temperature of 50 °F. Topography was diversified with valleys and well-developed flood plains.

The youngest flora from the Deschutes River Valley is the Deschutes flora dated at 4.3-5.3 m.y.b.p. (Evernden et al., 1964), latest Miocene to early Pliocene although earlier authors placed it in early to middle Pliocene (Chaney, 1938, 1959). The Deschutes flora is deposited in coarse tuffs suggesting a nearby volcano while diatomite beds indicate a lake basin. The flora is dominated by *Populus* (aspen) suggesting a fairly dry and cool climate. The presence here of *Salix* (willow), *Prunus* (cherry) and *Acer* (maple) indicate a stream border environment at a high elevation. An *Auchenia* (camel) fragment was found by Condon in association with this flora.

Pliocene floras are only questionably known from the State. Late Miocene floras from areas east of the Cascades show the increased effects of the rainshadow emerging in the Miocene as the Cascades are being emplaced during late Tertiary time. Floristic composition of these latest Miocene units is not altogether different from extant floras. By late Miocene/earl Pliocene time the general topography of the Oregon area we know today was established. Estimates for the rainfall of the northern Great Basin and Columbia Plateau at this time are around 15 to 17 inches per year, clearly reflecting the rain shadow of the Cascades. Chaney (1956) has stated that in addition to the loss of the warm tropical species, conditions of preservation diminished in this interval following the emplacement of the High Cascades. The larger volcanic episodes of the middle Miocene were over. While there was residual volcanic activity in the Pliocene, the thick accumulations of volcanic material conducive to plant preservation were a thing of the past. As a consequence, we know less of the Pliocene than of any other epoch in the Oregon Tertiary since the Eocene.

Very little has been published on Pleistocene floras in the State. Sullivan (1970) has radiocarbon dated fossil wood and cones of *Picea sitchensis* (spruce) from Cape Fisheries near Port Orford at 3500 years. Wolfe (1969) mentions an unpublished Pleistocene

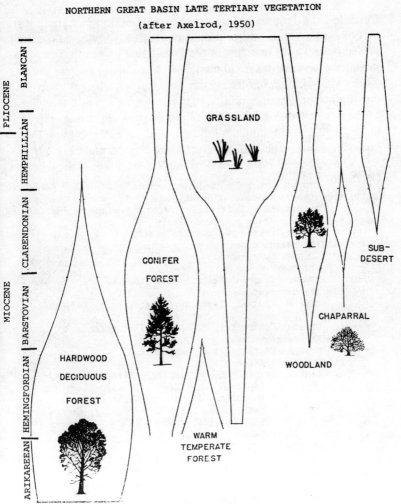

NORTHERN GREAT BASIN LATE TERTIARY VEGETATION

(after Axelrod, 1950)

PLIOCENE — BLANCAN | HEMPHILLIAN

MIOCENE — CLARENDONIAN | BARSTOVIAN | HEMINGFORDIAN | ARIKAREEAN

GRASSLAND

CONIFER FOREST

SUB-DESERT

CHAPARRAL

HARDWOOD DECIDUOUS FOREST

WOODLAND

WARM TEMPERATE FOREST

flora from Cape Blanco which is striking for its limited amounts of *Pseudotsuga* (Douglas fir) compared to rich late Pleistocene occurrences of this genus elsewhere in the Pacific Northwest. This distribution has been used by Wolfe to surmise that the extant dominance of Douglas fir did not take place here until quite late in the Pleistocene.

Baldwin (1950, 1973) notes the occurrences of unidentified fossil wood in the Pleistocene Coquille Formation exposed at Newport and Coos Bay. The wood is found as stumps and logs and has been C14 dated at 50,000 years. Because this is very near the limit for the radiocarbon method, these logs could be much older.

The following classification contains Oregon fossil plant genera listed in the published literature. Taxonomic revision has not been attempted.

ORDER: Thallophyta (spore plants)
FAMILY: Chlorphyceae (algae)
GENERA: *Botryococcus***, *Pediastrum***, *Tetraedon***
FAMILY: Aspergillaceae (fungi)
GENERA: *Cryptocolax*
FAMILY: Stictaceae (lichen)
GENERA *Lobaria*
ORDER: Bryophyta (liverwort)
GENERA: *Marchantites*
ORDER: Lycopodophyta
GENERA: *Lepidodendroid*, *Lycopodium*, *Selaginella***, *Stigmaria*
ORDER: Sphenophyta (ancestral joint grass)
GENERA: *Asterophyllites*, *Equisetum*, *Mesocalamites*, *Phyllotheca*
ORDER: Filicophyta (fern and fern-like)
GENERA: *Dicranophyllum*, *Pecopteris*
FAMILY: Marattiaceae
GENERA: *Danaeopsis*
FAMILY: Osmundaceae
GENERA: *Cladophlebis*, *Osmunda**, *Osmundites*
FAMILY: Schizaeceae
GENERA: *Aneimia*, *Lygodium*, *Schizopteris*, *Ruffordia*
FAMILY: Dicksoniaceae (Cyatheaceae)
GENERA: *Conipteris*, *Ticksonia*, *Thyrsopteris*
FAMILY: Dipteridinae
GENERA: *Clathropteris*, *Hausmannia*
FAMILY: Polypodiaceae
GENERA: *Acrostichum*, *Adiantites*, *Allantodiopsis*, *Asplenum*, *Davillia***, *Dennstaedtia*, *Dryopteris* (*Lastrea*), *Nephrolepis*, *Onchiopsis*, *Polypodium*, *Pteris*, *Pteridium*, *Scleropteris*, *Woodwardia**
FAMILY: Tempskyaceae
GENERA: *Tempskya*

GYMNOSPERMS
ORDER: Pteridospermophyta (seed ferns, extinct)
FAMILY: Claytoniaceae
GENERA: *Sagenopteris*
ORDER: Cycadophyta (cycad and cycad-like)
GENERA: *Ctenis*, *Ctenophyllus*, *Cycadeospermum*, *Dioon*, *Encephalartopsis*, *Macrotaeniopteris*, *Nilssonia*, *Pterophyllum*, *Ptilozamites*, *Taeniopteris*, *Williamsonia*
ORDER: Ginkgophyta
FAMILY: Ginkgoaceae
GENERA: *Ginkgo*, *Phoenicopsis*

*plant, spore or pollen
**spore or pollen only

ORDER: Coniferophyta (conifers)
      GENERA: *Cyclopitys, Podozamites, Taxites*
   FAMILY: Araucariaceae
      GENERA: *Araucaria*
   FAMILY: Cupressaceae
      GENERA: *Brachyphyllum, Chamaecyparis, Cupressus,*
             Juniperus*, Libocedrus
   FAMILY: Taxodiaceae
      GENERA: *Cunninghamia, Glyptostrobus*, Metasequoia*,*
             Sequoia*, Taxodium*
   FAMILY: Pinaceae
      GENERA: *Abies*, Keteleeria*, Cedrus**, Larix,*
             Picea*, Pinus*, Pseudotsuga*, Sphenolepidium,
             Thuja, Tsuga*
   FAMILY: Taxaceae
      GENERA: *Podocarpus**, Taxus, Torreya*
   FAMILY: Cephalotaxaceae
      GENERA: *Cephalotaxus*
   FAMILY: Gnetaceae
      GENERA: *Ephedra***

ANGIOSPERMS

ORDER: Salicales
   FAMILY: Salicaceae (willow)
      GENERA: *Populus*, Salix**
ORDER: Myricales
   FAMILY: Myricaceae (sweet gale)
      GENERA: *Comptonia, Myrica**
ORDER: Leitneriales
   FAMILY: Leitneriaceae (cork-wood)
      GENERA: *Leitneria*
ORDER: Juglandales
   FAMILY: Juglandaceae (walnut)
      GENERA: *Carya (Hicoria)*, Engelhardia, Juglans**
             Pterocarya*, Platycarya**
ORDER: Fagales
   FAMILY: Betulaceae (birch)
      GENERA: *Alnus*, Betula*, Carpinus*, Corylus*,*
             Ostrya*
   FAMILY: Fagaceae (beech)
      GENERA: *Castanea*, Castanopsis, Fagus*, Lithocarpus*,*
             Quercus*
ORDER: Urticales
   FAMILY: Ulmaceae (elm)
      GENERA: *Celtis*, Planera, Ulmus*, Zelkova**
   FAMILY: Moraceae (mulberry)
      GENERA: *Ficus, Maclura*
ORDER: Santales
   FAMILY Olacaceae
      GENERA: *Schoepfia***
ORDER: Aristolochiales
   FAMILY: Aristolochiaceae (birthwort, ginger)
      GENERA: *Aristolochina, Asarum*

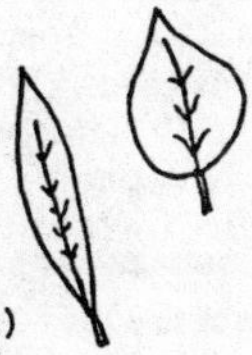

*plant, spore or pollen
**spore or pollen only

ORDER: Ranales
    FAMILY: Nymphaceae (water lily)
      GENERA: *Nymphaea, Nymphaeites*
    FAMILY: Trochodendraceae
      GENERA: *Euptelea, Trochodendron*
    FAMILY: Cercidiphyllaceae
      GENERA: *Cercidiphyllum*
    FAMILY: Ranunculaceae
      GENERA: *Clematis*
    FAMILY: Berberidaceae (barberry)
      GENERA: *Berberis, Mahonia*
    FAMILY: Menispermaceae (moonseed)
      GENERA: *Abuta, Chandlera, Cissampelos, Cocculus, Diploclisia, Hyperbaena, Hyserpa, Odontocaryoidea*
    FAMILY: Magnoliaceae (magnolia)
      GENERA: *Drimys, Liriodendron, Magnolia*
    FAMILY: Anonaceae (custard-apple)
      GENERA: *Anona, Artototrys, Asimina, Fississtigma, Polyalthia*
    FAMILY: Monimiaceae
      GENERA: *Siparuna*
    FAMILY: Lauraceae (laurel)
      GENERA: *Cinnamomum, Cryptocarya, Laurocarpus, Laurophyllum, Laurus, Lindera, Litsea, Nachilus, Nectandra, Ocotea, Oreodaphne, Persea,*
ORDER: Rosales   *Phoebe, Sassafrass, Umbellularia*
    FAMILY: Saxifragaceae
      GENERA: *Hydrangea, Philadelphus, Ribes*
    FAMILY: Hamamelidaceae (witch hazel)
      GENERA: *Exbuchlandia, Hamamelis, Liquidambar**
    FAMILY: Platanaceae (plane tree)
      GENERA: *Platanophyllum, Platanus**
    FAMILY: Rosaceae (rose)
      GENERA: *Amelanchier, Cercocarpus, Chrysobalanus, Crataegus, Holodiscus, Photinia, Prunus, Pyrus, Rosa, Sorbus, Spiraea*
    FAMILY: Leguminosae (pea)
      GENERA: *Acacia, Albizzia, Amorpha, Astragalus***, *Cassia, Caesalpina***, *Cercis, Cladrastis, Gleditsia, Gymnocladus, Inga, Leguminosites, Lonchocarpus, Micropodium, Sophora**
ORDER: Geraniales
    FAMILY: Rutaceae (rue)
      GENERA: *Amyris, Evodia, Ptelea*
    FAMILY: Simarubaceae
      GENERA: *Ailanthus*
    FAMILY: Meliaceae (mahogany)
      GENERA: *Cedrela, Entandrophragma, Walsura*
    FAMILY: Malpighiaceae
      GENERA: *Bunchosia, Hiraea*

*plant, spore or pollen
**spore or pollen only

ORDER: Geraniales
      FAMILY: Polygalaceae (milkwort)
         GENERA: *Securidaca***, *Xanthophyllum***
      FAMILY: Euphorbiaceae (spurge)
         GENERA: *Alchornea, Aporosa, Drypetes, Mallotus, Sapium*
ORDER: Sapindales
      FAMILY: Buxaceae (box)
         GENERA: *Pachysandra***
      FAMILY: Anacardiaceae (cashew)
         GENERA: *Astronium, Rhus, Tapirira*
      FAMILY: Aquifoliaceae (holly)
         GENERA: *Ilex**
      FAMILY: Celastraceae (staff-tree)
         GENERA: *Celastrus, Euonymous, Tripterygium*
      FAMILY: Icacinaceae (liana)
         GENERA: *Palaeophytocrene*
      FAMILY: Aceraceae (maple)
         GENERA: *Acer*, Dipteronia*
      FAMILY: Hippocastanaceae (horse chestnut)
         GENERA: *Aesculus*
      FAMILY: Staphyleaceae (bladder nut)
         GENERA: *Staphylea*
      FAMILY: Sapindaceae (soapberry)
         GENERA: *Allophylus, Cupania, Sapindus, Thouinia*
      FAMILY: Sabiaceae
         GENERA: *Meliosma*
ORDER: Rhamnales
      FAMILY: Rhamnaceae (buckthorn)
         GENERA: *Berchemia, Ceanothus, Colubrina, Gouania, Rhamnus**
      FAMILY: Vitaceae
         GENERA: *Parthenocissus, Tetrastigma, Vitis**
ORDER: Malvales
      FAMILY: Elaeocarpaceae
         GENERA: *Sloanea*
      FAMILY: Tiliaceae (linden)
         GENERA: *Grewia, Grewiopsis, Tilia**
      FAMILY: Malvaceae (mallow)
         GENERA: *Anoda, Gossypium, Urena*
      FAMILY: Sterculiaceae (sterculia)
         GENERA: *Pterospermum, Sterculia*
ORDER: Parietales
      FAMILY: Dilleniaceae
         GENERA: *Dillenites, Saurauia, Tetracera*
      FAMILY: Actinidiaceae
         GENERA: *Actinidia*
      FAMILY: Theacea (tea)
         GENERA: *Gordonia, Schima*
      FAMILY: Flacourtiaceae
         GENERA: *Casearia, Erythrospernum, Idesia, Xylosma*

*plant, spore or pollen
**spore or pollen only

```
ORDER: Myrtales
    FAMILY: Elaeagnaceae (oleaster)
        GENERA: Elaeagnus**, Shepherdia**
    FAMILY: Lythraceae (loosestrife)
        GENERA: Lagerstroemia**
    FAMILY: Nyssaceae
        GENERA: Alangiophyllum, Nyssa*, Palaeonyssa
    FAMILY: Alangiaceae
        GENERA: Alangium
    FAMILY: Combretaceae
        GENERA: Terminalia
    FAMILY: Myrtaceae (myrtle)
        GENERA: Calyptranthes
    FAMILY: Hydrocaryaceae (Onagraceae)
        GENERA: Trapa, Xylonagra
ORDER: Umbellales
    FAMILY: Araliaceae
        GENERA: Aralia, Oreopanax
    FAMILY: Cornaceae (dogwood)
        GENERA: Cornus, Mastixioidiocarpum
ORDER: Ericales
    FAMILY: Clethraceae
        GENERA: Clethra
    FAMILY: Ericaceae
        GENERA: Andromeda, Arbutus, Rhododendron, Vaccinium*
ORDER: Primulales
    FAMILY: Myrsinaceae
        GENERA: Reptonia
ORDER: Ebenales
    FAMILY: Sapotaceae (sapondilla)
        GENERA: Chrysophyllum, Lucuma
    FAMILY: Ebenaceae
        GENERA: Diospyros
    FAMILY: Symplocaceae
        GENERA: Symplocos
    FAMILY: Styracaceae (storax)
        GENERA: Alniphyllum, Halesia*
ORDER: Contortae
    FAMILY: Oleaceae (olive)
        GENERA: Fraxinus, Osmanthus
    FAMILY: Loganiaceae
        GENERA: Strychnos
    FAMILY: Gentianaceae
        GENERA: Nymphoides
    FAMILY: Apocyanaceae (dogbane)
        GENERA: Apocynum, Forsteronia, Tabernaemontana
    FAMILY: Asclepiadaceae (milkweed)
        GENERA: Vincetoxicum
ORDER: Tubiflorae
    FAMILY: Verbenaceae (verain)
        GENERA: Cornuita**, Holmskioldia, Porana
```

*plant, spore or pollen
**spore or pollen only

ORDER: Tubiflorae
    FAMILY: Scrophulariaceae
        GENERA: *Paulownia*
    FAMILY: Bignoniaceae
        GENERA: *Callichlamys, Catalpa, Cordia*
ORDER: Rubiales
    FAMILY: Rubiaceae (madder)
        GENERA: *Galium***, *Psychotria*
    FAMILY: Caprifoliaceae (honeysuckle)
        GENERA: *Symphoricarpos, Viburnum*
    FAMILY: Valerianaceae
        GENERA: *Valeriana***
    FAMILY: Compositae
        GENERA: *Ambrosia***, *Saussurea*
ORDER: Pandanales
    FAMILY: Typhaceae (cat-tail)
        GENERA: *Typha**
    FAMILY: Sparganiaceae (bur-reed)
        GENERA: *Sparangium**
ORDER: Helobiae
    FAMILY: Potamogetonaceae (pond weed)
        GENERA: *Potamogeton**
    FAMILY: Gramineae (grass)
        GENERA: *Bambusa*
    FAMILY: Cyperaceae (grass)
        GENERA: *Cyperacites*
ORDER: Principes
    FAMILY: Palmae (palm)
        GENERA: *Attalea, Palmoxylon, Sabalites*
ORDER: Arales
    FAMILY: Araceae (skunk cabbage)
        GENERA: *Lysichiton*
ORDER: Liliales
    FAMILY: Liliaceae
        GENERA: *Smilax, Yuccites*

Carpolithus-winged conifer seed
Cordaianthus-fruit of conifer
Samaropsis-seed-like

*plant, spore or pollen
**spore or pollen only

51

OREGON FOSSIL PLANT BEARING FORMATIONS

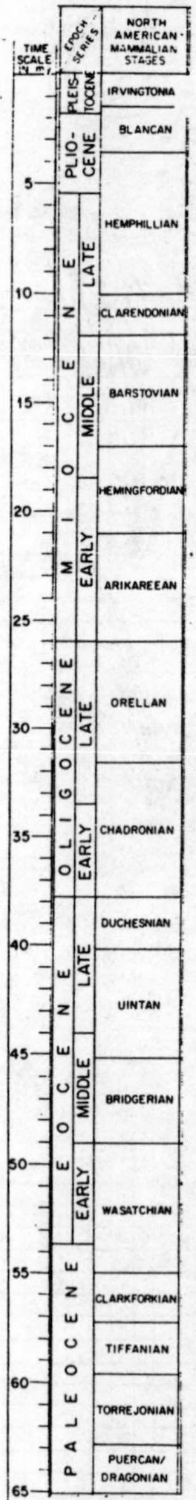
TIME SCALE IN m.y.
EPOCH SERIES
NORTH AMERICAN MAMMALIAN STAGES
PLEISTOCENE
IRVINGTONIAN
PLIOCENE
BLANCAN
MIOCENE
LATE
HEMPHILLIAN
CLARENDONIAN
MIDDLE
BARSTOVIAN
HEMINGFORDIAN
EARLY
ARIKAREEAN
OLIGOCENE
LATE
ORELLAN
EARLY
CHADRONIAN
EOCENE
LATE
DUCHESNIAN
UINTAN
MIDDLE
BRIDGERIAN
EARLY
WASATCHIAN
PALEOCENE
CLARKFORKIAN
TIFFANIAN
TORREJONIAN
PUERCAN/ DRAGONIAN
5
10
15
20
25
30
35
40
45
50
55
60
65

Coquille Fm.

Deschutes flora 4.3-5.3m.y.b.p. Deschutes Fm.
Troutdale flora Troutdale Fm.
The Dalles flora The Dalles Fm.
Rattlesnake flora 6.4 m.y.b.p. Rattlesnake Fm.

Sardine Fm. (Little Butte Volcanics)
Stinking Water flora 12.1 m.y.b.p. Columbia River Group
Molalla flora 12.2-12.9 m.y.b.p. Little Butte Volcanics (Molalla Fm.)
Sparta flora 13 m.y.b.p., Blue Mountain flora 13 Columbia River Group
Trout Creek flora 13.1 m.y.b.p. Trout Creek Fm.
Collawash flora 13-16 m.y.b.p. Little Butte Volcanics (?Eagle Creek Fm.
Eagle Creek flora 13-16 m.y.b.p. Little Butte Volcanics (Eagle Creek Fm
Mascall flora 15.4 m.y.b.p. Mascall Fm.
Sucker Creek flora Rockville flora 16.7 m.y.b.p. Sucker Creek Fm.
Alvord Creek flora 21.3 m.y.b.p. Alvord Creek Fm.

Thomas/Bilyeu Creek flora ?26.m.y.b.p. Little Butte Volcanics (Mehama Volcanics)
Scio flora Little Butte Volcanics ("Scio Beds")
Lyons flora Little Butte Volcanics
Coos Bay Tunnel Point Fm.
Bridge Creek flora 31.5 m.y.b.p. John Day Fm. (Bridge Creek Shales)
Willamette Junction Goshen flora Fisher Fm.
Rujada flora
Clarno Nut Beds 34.0 m.y.b.p. Clarno Fm.
Sweet Home flora Little Butte Volcanics (Eugene Fm.)
Coos Bay Bastendorff Fm.

Clarno flora 41.0-43.1 m.y.b.p. Clarno Fm.
Comstock flora Fisher Fm. (Colestin Fm.)
Coos Bay flora Coaledo Fm.

Rickreall Limestone Member Yamhill Fm.

*m.y.b.p. Radiometric date in millions of years before present

135	CRETACEOUS	Elk River Curry Co. Humbug Mtn. Tempskya Baker Co.
180	JURASSIC	Riddle . Douglas Co. Myrtle Group (Riddle Fm.)
225	TRIASSIC	
270	PERMIAN	
315	PENNSYLVANIAN	Spotted Ridge flora Spotted Ridge Fm.

The following are fossil plant ranges listed in the published literature.

	Spotted Ridge flora	Jurassic, S.E. Ore.	Cretaceous, Cent. Or.	Clarno flora	Coos Bay flora/pol.	Comstock flora	Clarno Nut Beds	Sweet Home flora	Bridge Creek flora	Goshen-Will. Junct.	Rujada flora	Lyons flora	Scio flora	Bilyeu-Thomas Cr.	Alvord Creek flora	Sucker Creek flora	Rockville flora	Mascall flora	Sucker-Trout Cr. pol	Trout Cr. flora	Eagle Cr. flora	Molalla-Coll. pollen	Collawash flora	Weyerh.-Old Molalla	Masc.-B.Mtn.-St.W. pol	Blue Mtn. flora	Sparta-Keating flora	Stinking Water flora	Rattlesnake flora	The Dalles flora	Troutdale flora	Deschutes flora
Abies (fir)								●	●								●	●	△	●		△			△	●	▲	●		●		
Acacia (acacia)		●																														
Acer (maple)						●		●	●	●	●	●		●	●	●	●	●	△	●	●	△	●	●		●	▲			●	●	●
Acrostichum (fern)				●				●																								
Actinidia								●																								
Adiantites	●																															
Aesculus (buckeye)								●																								
Ailanthus (tree of heaven)							●	●																						●		
Alangium				●				●		●	●																					
Alangiophyllum				●																												
Alchornea				●																												
Albizzia														●																		
Allophylus				●																												
Allantodiopsis (fern)			●		●																											
Alnus (alder)			△		●	●	●	●	●	●						●	△	●	●	●			△	●	●	●						
Ambrosia																∶		△														
Amelanchier (serviceberry)						●					●		●	●					●					●				●				
Amorpha (indigo bush)													●																		●	●
Amyris													●																			
Andromeda (bog rosemary)				●																												
Aneimia (fern)				●																												
Anoda													●																			
Anona			●		●			●	●																							
Apocynum (dogbane)													●					●														
Aporosa				●																												
Aralia (aralia)			●	●				●																								
Araucarites (conifer)	●																															
Arbutus (madrone)							●	●								●		●			●			●	●		●					
Aristolochia (birthwort)			●					●																								
Artototrys																																
Asarum (ginger)								●																								
Asimia			△																													
Asplenium (fern)			●																													
Asterophyllites	●																															
Astragalus																								△								
Astronium			●	●				●	●																							
Attalea (palm nut)																																
Bambusium (bamboo)																		●														
Berberis (barberry)																		●														
Berchemia																						●	●									
Betula (birch)			△		●	●		●								●	●	●	△			△	●	●	△	●	▲					
Botryococcus algae																●			●													
Brachyphyllum (conifer)	●																															
Bunchosia											●																					
Caesalpinia																								△								
Callichlamus				●							●																					
Calyptranthes				●				●																								
Carpalites (seed)																		●														
Carpinus (hornbeam)			△		●	●	●										●		△	●	●							●				
Carpolithus (seed-like)																							●									
Carya (hickory)			△		●	●	●	●								●	●	●	△	●			△	●	△		▲					
Casearia																																
Cassia (senna)								●																								

	Sptd. R.	Jur.	Cret.	Clarno	Coos B.	Comst.	Cl. Nut	Swr. Hme.	Br. Cr.	Goshn.-Wil.	Ruj.	Lyons	Scio	Bily-Th.	Alvord	Sr. Cr.	Rkv.	Masc.	Sr.-Tr. Cr.	Tr. Cr.	Eagle Cr.	Mol.-Coll.	Coll.	Wey.-Old M.	Masc.-B. Mt.	Blue Mtn.	Sparta	Stinkg. W.	Rattle	Dalles	Troutd.	Desch.
Castanea (chestnut)				△												●	●	△	●	●	●	△										
Castanopsis (chinquapin)						●	●		●	●		●		●				●						●		●			●			
Catalpa				●	●		●																									
Ceanothus (buckbrush)							●						●																			
Cedrus (cedar)																													△			
Cedrela (cedar)				●								●	●									●		●		●						
Celastrus			●		●		●																									
Celtis (hackberry)																△	●						△		△							
Cephalotaxus (yew)																																
Cercidiphyllum (katsura)				●	●	●			●			●																				
Cercis (redbud)					●	●																●		●					●			
Cercocarpus (mtn. mahogany)																																
Chamaecyparis (false cedar)																																
Chandlera				●																												
Chrysobalanus (cocoa plum)																																
Chrysophyllum																																
Cinnamomum (camphorwood)			●	●		●	●	●			●			●								●										
Cissampelos																																
Cladrastis (yellowwood)																																
Cladophlebis (fern)	●																															
Clematis																●																
Clethra (white alder)								●																								
Cocculus																							●	●								
Colubrina					●																											
Comptonia (fern)																																
Cordaianthus	●																															
Cordia (geiger tree)			●	●		●		●																		●						
Cornus (dogwood)				●		●	●	●	●			●			●	●					△	●		●					●		●	●
Cornutia																						△										
Corylus (hazel nut)				△		●	●																		△							
Crataegus (black hawthorn)						●	●	●				●			●		●	●		●	●		●	●	●		●					
Cryptocarya			●	●		●																										
Cryptocolax (fungi)			●																													
Ctenis	●																															
Ctenophyllum	●	●																														
Cunninghamia						●	●	●														●	●	●								
Cupania						●																										
Cupressus (cypress)																																
Cycadeospermum	●																															
Cyclopitys (conifer)	●																															
Cyperacites (grass)						●		●								●					●						●			●		●
Danaeopsis (fern)	●																															
Davallia (fern)																			△													
Dennstaedtia (fern)			●									●																				
Dicksonia (fern)	●																															
Dicranophyllum	●																															
Dillenites			●				●			●																						
Dioon (fern)			●																													
Diospyros (persimmon)			●	●		●	●	●					●			●		●				●										
Diploclisia																																
Dipteronia						●		●																								
Drimys																●																
Dryopteris (fern)				●														●														
Drypetes																																
Elaeagnus																		△														
Encephalartopis	●																															
Engelhardia							●	●	●			●												●								
Entandrophragma					●																											
Ephedra (Mormon tea)				△														△						△		△			●			
Equisetum (joint-grass)			●	●				●	●	●		●	●		●	●		●		●												
Erythrospernum																																
Euonymus (burning bush)																								●								
Euptela			●																													
Evodia			●																													

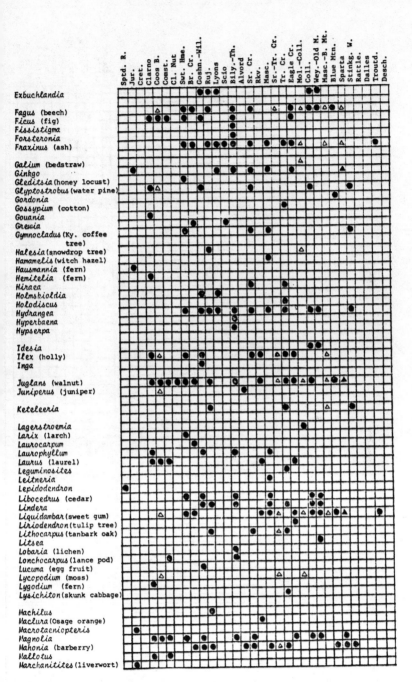

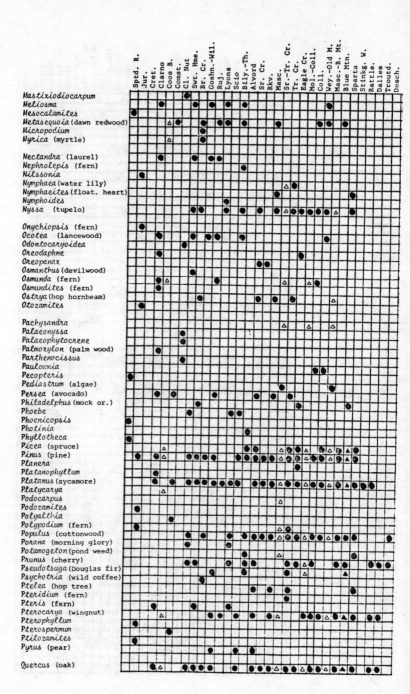

	Sptd. R.	Jur.	Cret.	Clarno	Coos B.	Comst.	Cl. Nut	Swt. Hme.	Br. Cr.	Goshn.-Wil.	Ruj.	Lyons	Scio	Bily.-Th.	Alvord	Sr. Cr.	Rkv.	Masc.	Sr.-Tr. Cr.	Tr. Cr.	Eagle Cr.	Mol.-Coll.	Coll.	Wey.-Old M.	Masc.-B. Mt.	Blue Mtn.	Sparta	Stinkg. W.	Rattle.	Dalles	Troutd.	Desch.
Reptonia (gargura)							●						●																			
Rhamnus (cascara)				●	●	●			●	●														△					●			●
Rhododendron					●	●										●					●	● ●										
Rhus (sumac)					●	●			●		●			●	● ●		●															
Ribes (currant)									●				●															●				
Rosa							●		●			●				●												●				
Ruffordia (fern)	●																															
Sabalites (palm leaf)		●								●											●											
Sagenopteris (fern)	●																															
Salix (willow)			●	△			●	●		●	●		●		●			△	●	●	△	● ●	△	● ●	●	●	●	●	●	●	●	●
Samaropsis	●																															
Sapindus (soap berry)																																
Sapium							●																									
Sassafrass				●		●				●				●	●						●		●									
Saurauia																																
Saussurea															●																	
Schima	●																															
Schizopteris				●																												
Schoepfia																					●											
Scleropteris	●																															
Securidaca (milkwort)																△																
Sequoia (redwood)				●	●		●	●	●	●	●														● ●		● ●					
Shepherdia															△																	
Siparuna				●				●																								
Sloanea							●		●		●					●							●					●				
Smilax							●		●		●			●		●			●	● ●		●										
Sophora															●																	
Sorbus (mtn. ash)															●								●									
Sparangium			△																													
Sphenolepidium	●																											●				
Spiraea																					●											
Staphylea (bladder nut)					●																●											
Sterculia	●																															
Stigmaria	●																															
Strychnos				●																												
Symplocos (sweet leaf)				●																												
Symphoricarpos (snowberry)													●																			
Tabernaemontana				●																												
Taeniopteris	●																															
Tapirira				●																												
Taxodium (bald cypress)			△							●				●							●	●		●								
Taxus (yew)		●																														
Tempskya	●																															
Terminalia								● ●																								
Tetracera (liana vine)			●			●			●		●																					
Tetraedon (algae)																							●									
Tetrastigma				●																												
Thouinia (soap berry)																							●									
Thuja (arborvita)																									● ●							
Thyrsopteris	●																															
Tilia (basswood)			△					●	● ●	●	●	●						△	●	● ●				△		△						
Torreya (conifer)						●																			●							
Trapa (water ches.)																									●							
Tripterygium									●																							
Trochodendron				●																												
Trochodendroxylon				●																												
Tsuga (hemlock)							●			●									△ ●		△			△ ●	△							
Typha (cattail)			△		●						● ●	●	●	△ ●		● ●			△							△		△				
Ulmus (elm)			△		●	●		●			●	● ●		△	●	●	△	● ●	△					△		● ● ●			●			
Umbellularia (laurel)				● ●						●		●												●								●
Urena													●																			
Vaccinium (blueberry)							●									△					●	△				●						

Column headers (left to right):

1. Spotted Ridge flora
2. Jurassic, S.E. Ore.
3. Cretaceous, Cent. Ore.
4. Clarno flora
5. Coos Bay flora/pollen
6. Comstock flora
7. Clarno Nut Beds
8. Sweet Home flora
9. Bridge Creek flora
10. Goshen-Will. Junct.
11. Rujada flora
12. Lyons flora
13. Scio flora
14. Bilyeu-Thomas Cr.
15. Alvord Creek flora
16. Sucker Creek flora
17. Rockville flora
18. Mascall flora
19. Sucker-Trout Cr. pollen
20. Trout Cr. flora
21. Eagle Cr. flora
22. Molalla-Coll. pollen
23. Collawash flora
24. Weyerh.-Old Mollala
25. Masc.-B.Mtn.-St.W. pollen
26. Blue Mtn. flora
27. Sparta-Keating flora
28. Stinking Water flora
29. Rattlesnake flora
30. The Dalles flora
31. Troutdale flora
32. Deschutes flora

Taxon	1	2	3	4	5	6	7	8	9	10	11	12	13	14	15	16	17	18	19	20	21	22	23	24	25	26	27	28	29	30	31	32
Valeriana									●																	●						
Viburnum									●	●	●																●					
Vincetoxicum																																
Vitis (grape)				●				●					●								●	△	●	●							●	
Walsura									●																							
Williamsonia	●																															
Woodwardia (fern)				●									●	●					△													
Xanthophyllum (milkwort)																							△									
Xylonagra (water ches.)																							△									
Xylosma																							●	●								
Yuccites	●																															
Zelkova (keaki tree)										●												△	●	●		●						●
Addendum:																																
Abuta			●																													
Alniphyllum							●						●																			
Conipteris		●																														
Selaginella																									△							
Taxites	●																															

Legend:

● megaflora
▲ pollen and megaflora
△ pollen

OREGON FOSSIL PLANT LOCALITIES

Following is a list of Oregon fossil plant localities. The reference numbers are to authors listed in the bibliography who have published on the specific localities.

PENNSYLVANIAN
1. Spotted Ridge, @ 10 mi. SW of Suplee (Mills Ranch), Crook Co. Spotted Ridge Fm. Spotted Ridge flora. Refs. 8c, 119, 170.

JURASSIC
2. Buck Peak, 10 mi. NW of Riddle, Douglas Co. Riddle Fm. Refs. 49f, 61, 103b.
3. Cow Creek (Nichols Station), SW of Riddle, Douglas Co. Riddle Fm. Refs. 49f, 61, 103b.
4. Riddle, Douglas Co. Riddle Fm. Refs. 49f, 61, 103b.
5. Thompson Creek, NW of Riddle, Douglas Co. Riddle Fm. Refs. 49f, 61, 103b.

CRETACEOUS
6. Elk River, Curry Co. Humbug Mtn. Conglomerate. Refs. 104, 114.
7. Lightning Creek, 5 mi. N of Sumpter, Baker Co. (Greenhorn area). No Fm. Refs. 42, 169.
8. Mitchell, Wheeler Co. No Fm. Refs. 69a.

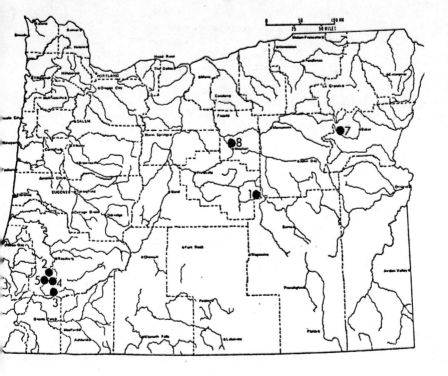

EOCENE PLANTS

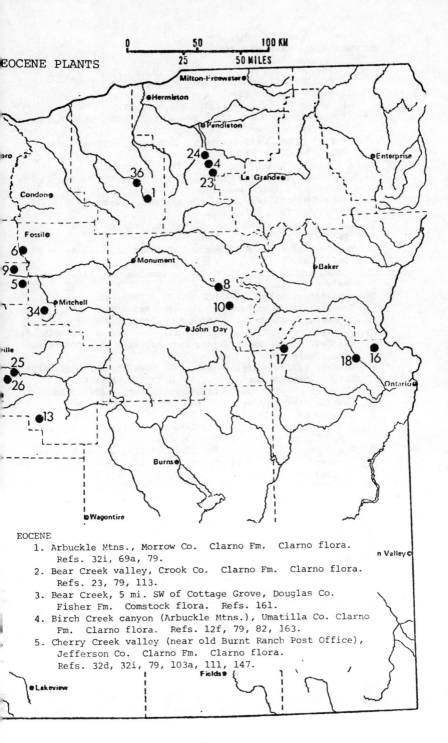

EOCENE PLANTS

EOCENE

1. Arbuckle Mtns., Morrow Co. Clarno Fm. Clarno flora.
   Refs. 32i, 69a, 79.
2. Bear Creek valley, Crook Co. Clarno Fm. Clarno flora.
   Refs. 23, 79, 113.
3. Bear Creek, 5 mi. SW of Cottage Grove, Douglas Co.
   Fisher Fm. Comstock flora. Refs. 161.
4. Birch Creek canyon (Arbuckle Mtns.), Umatilla Co. Clarno
   Fm. Clarno flora. Refs. 12f, 79, 82, 163.
5. Cherry Creek valley (near old Burnt Ranch Post Office),
   Jefferson Co. Clarno Fm. Clarno flora.
   Refs. 32d, 32i, 79, 103a, 111, 147.

61

6. Clarno, Wheeler Co. Clarno Fm. Clarno flora.
   Refs. 8b, 8d, 32d, 32i, 69a, 79, 162, 183.
7. Comstock, 5 mi SW of Cottage Grove, Douglas Co. Fisher
   Fm. Comstock flora. Refs. 161, 83, 177.
8. Crockett Knob, 5 mi. SE of Susanville, Grant Co. Clarno
   Fm. Clarno flora. Refs. 12f, 141.
9. Currant Creek, Jefferson Co. Clarno Fm. Clarno flora.
   Refs. 79, 103a, 147.
10. Dixie Mtn., Grant Co. Clarno Fm. Refs. 69a.
11. Elk Creek-Coon Creek valley, 15 mi. SE of Drain, Douglas Co.
    Fisher Fm. Refs. 161, 83.
12. Evans Creek valley, 25 mi. NE of Grants Pass, Jackson Co.
    ?Colestin Fm. Refs. 161.
13. Hampton Butte, Crook Co. Clarno Fm. Clarno flora.
    Refs. 23, 79, 113.
14. Hanson Coal Mine, N of Medford, Jackson Co. ?Colestin Fm.
    Refs. 161.
15. Hobart Butte, 5 mi. SE of Drain, Douglas Co. Fisher Fm.
    Refs. 83, 161.
16. Huntington, Malheur Co. Clarno Fm. Refs. 69a.
17. Ironside Mtn., Malheur Co. No Fm. Refs. 69a.
18. Jamieson, Malheur Co. Clarno Fm. Refs. 69a.
19. Little River, SE of Glide, Douglas Co. Fisher Fm. Refs. 161.
20. Meadows, 20 mi. NE of Grants Pass, Jackson Co. ?Colestin
    Fm. Refs. 161.
21. Nehalem River valley, Columbia Co. Cowlitz Fm. Refs. 229.
22. Oregon Portland Cement Co., SW of Dallas, Polk Co.
    Yamhill Fm. (Rickreall Limestone Member). Refs. 12c.
23. Pearson Creek valley, S of Pendleton, Umatilla Co.
    Clarno Fm. Refs. 69a, 82, 163.
24. Pilot Rock, 10 mi. S of Pendleton, Umatilla Co. Clarno
    Fm. Clarno flora. Refs. 32i, 69a, 79, 82.
25. Post, Crook Co. Clarno Fm. Clarno flora. Refs. 8a, 79.
26. Riverside Ranch, 18 mi. SE of Prineville, Crook Co.
    Clarno Fm. Clarno flora. Refs. 32d, 32i, 79.
27. Riverton, Coos Co. Coaledo Fm. Coos Bay flora.
    Refs. 3, 49b.
28. Sams valley, 15 mi. NE of Grants Pass, Jackson Co.
    ?Colestin Fm. Refs. 69a, 111, 161.
29. Scotts valley, 5 mi. SE of Drain, Douglas Co. Fisher
    Fm. Refs. 161.
30. Siskiyou Summit, 10 mi. SE of Ashland, Jackson Co.
    Colestin Fm. Refs. 161.
31. Table Rock, 15 mi. E of Grants Pass, Jackson Co.
    ?Colestin Fm. Refs. 161.
32. Timber, Washington Co., Cowlitz Fm. Refs. 229.
33. Van Dyke cliffs, 4 mi. N of Ashland, Jackson Co.
    ?Colestin Fm. Refs. 83, 161.
34. West Branch Creek, 4 mi. SW of Mitchell, Wheeler Co.
    Clarno Fm. Clarno flora. Refs. 32i, 79.
35. White Point, 10 mi. SE of Ashland, Jackson Co. Colestin
    Fm. Refs. 161.
36. Willow Creek valley, SE of Heppner, Morrow Co. Clarno
    Fm. Clarno flora. Refs. 79, 82, 129, 163.

EARLY OLIGOCENE FOSSIL PLANT LOCALITIES

1. Clarno Nut Beds, 2 mi. E of Clarno, Wheeler Co. Clarno
   Fm. Clarno flora. Refs. 127, 183.
2. Holley, Linn Co. Eugene Fm. Sweet Home flora. Refs.69, 174.
3. Mist, Columbia Co. Keasey Fm. Refs. 142b, 247.
4. Sweet Home, Linn Co. Eugene Fm. Sweet Home flora.
   Refs. 161, 174.

LATE OLIGOCENE FOSSIL PLANT LOCALITIES

5. Ashwood (5 mi. S of), Jefferson Co. John Day Fm.
   Refs. 27c, 161.
6. Allen Ranch, 9 mi. NW of Mitchell (became Wade Ranch),
   Wheeler Co. John Day Fm. Bridge Creek flora. Refs. 32d,
   32i.
7. Bilyeu Creek (Neal Creek), Linn Co. Mehama Volcanics
   (Little Butte Volcanics). Thomas Creek flora. Refs. 102, 161.
8. Bridge Creek valley, Wheeler Co. John Day Fm. Bridge Creek
   flora. Refs. 27, 27b, 32b, 32d, 32i, 103a, 147.
9. Burmester Creek valley, E of Albany, Linn Co. Little
   Butte Volcanics. Refs. 161.
10. Butte Creek valley (Scotts Mills), Clackamas Co. "Butte
    Creek Beds", (Little Butte Volcanics). Refs. 161, 201d.
11. Cants Ranch, 10 mi. W of Dayville (Butler Basin), Grant Co.
    John Day Fm. Bridge Creek flora. Refs. 32d, 32i, 103a.
12. Clarno (1 mi. NE of), Wheeler Co. John Day Fm. Bridge
    Creek flora. Refs. 27b, 32d, 32i, 210.
13. Coal Creek, 15 mi. NE of Salem, Marion Co. "Butte Creek
    Beds", (Little Butte Volcanics). Refs. 52.
14. Coburg Hills, Lane Co. Fisher Fm. (Little Butte Volcanics).
    Willamette Junction flora. Refs. 12f, 27b, 112, 161, 224.
15. Cove Creek, 5 mi. NE of Clarno, Wheeler Co. John Day Fm.
    Bridge Creek flora. Refs. 32d, 210.
16. Crabtree Creek, NE of Lebanon, Linn Co. Little Butte
    Volcanics. Refs. 161.
17. Crooked River valley, Crook Co. John Day Fm. Bridge
    Creek flora. Refs. 32b, 32d, 32i.
18. Dugout Gulch, 2 mi. NE of Clarno, Wheeler Co. John Day
    Fm. Bridge Creek flora. Refs. 32d, 32i, 162, 210.
19. Fossil (at schoolhouse), Wheeler Co. John Day Fm.
    Bridge Creek flora. Refs. 27b, 103a.
20. Franklin Butte, Linn Co. "Scio Fm." (Little Butte
    Volcanics). Scio flora. Refs. 177a.
21. Gawley Creek, 15 mi. SE of Silverton, Clackamas Co. Little
    Butte Volcanics. Refs. 161.
22. Goshen, Lane Co. Fisher Fm. (Little Butte Volcanics)
    Goshen flora. Refs. 27b, 34, 161, 224.
23. Gray Ranch (McCullough Ranch), 10 mi. E of Post, Crook Co.
    John Day Fm. Bridge Creek flora. Refs. 27b, 32b, 32d.
24. Grizzly Peak (Shale City), Jackson Co. Little Butte
    Volcanics. Refs. 27b, 103, 161.
25. Hayden Bridge, Springfield, Lane Co. Little Butte Volcanics.
    Refs. 161.
26. Jasper, Lane Co. Little Butte Volcanics. Refs. 161.
27. Kincaid Ranch, 5 mi. SE of Ashland, Jackson Co. Little
    Butte Volcanics. Refs. 103, 161.

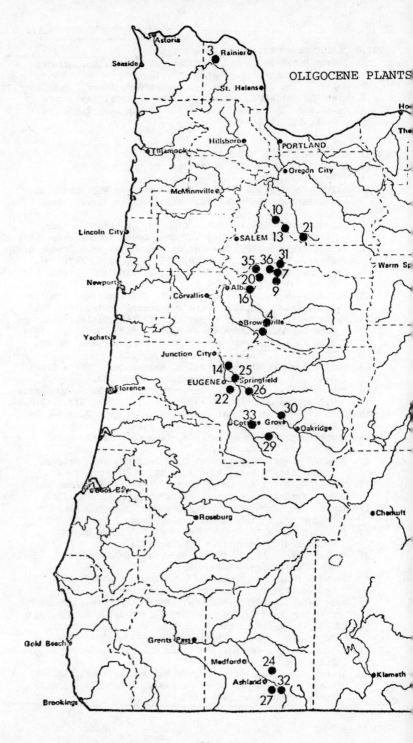

OLIGOCENE PLANTS

64

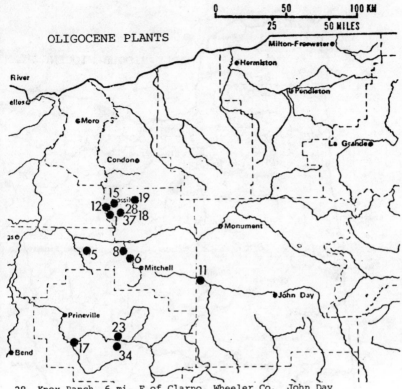

OLIGOCENE PLANTS

28. Knox Ranch, 6 mi. E of Clarno, Wheeler Co. John Day
    Fm. Bridge Creek flora. Refs. 8b, 27b, 162, 210.
29. Laying Creek, Lane Co. Little Butte Volcanics. Rujada
    flora. Refs. 108, 161.
30. Lookout Point Dam (Landax), Lane Co. Little Butte Volcanics.
    Refs. 27b, 161.
31. Lyons, Linn Co., Little Butte Volcanics. Lyons flora.
    Refs. 137.
32. Murphy Springs, SE of Ashland, Jackson Co. ?Little Butte
    Volcanics. Refs. 103.
33. Rat Creek, 3 mi. N of Dorena Dam, Lane Co. Little Butte
    Volcanics. Refs. 161.
34. Reams Ranch, 7 mi. SE of Post, Crook Co. John Day Fm.
    Bridge Creek flora. Refs. 32d.
35. Scio, Linn Co. "Scio Fm", (Little Butte Volcanics).
    Scio flora. Refs. 177.
36. Thomas Creek, 8 mi. E of Jordan, Linn Co. Mehama Volcanics
    (Little Butte Volcanics). Thomas Creek flora.
    Refs. 27b, 56, 56a, 102, 161.
37. Twickenham, Wheeler Co. John Day Fm. Bridge Creek flora.
    Refs. 32i.

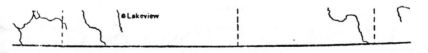

EARLY/MIDDLE MIOCENE PLANTS

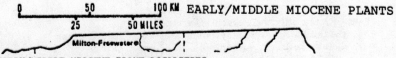

0   50   100 KM   EARLY/MIDDLE MIOCENE PLANTS

     25   50 MILES

EARLY/MIDDLE MIOCENE PLANT LOCALITIES

   1. Alvord Creek, Harney Co.   Alvord Creek Fm.   Alvord Creek
      flora.   Refs. 9, 32i.
   2. Austin (S of), Grant Co.   Columbia River Basalt.   Blue
      Mountain flora.   Refs. 32c, 33, 151.
   3. Bonneville (in railroad cut), Multnomah Co.   Eagle Creek
      Fm. (Little Butte Volcanics) Eagle Creek flora. Refs. 32a.
   4. Cascadia (5 mi. NE of), Linn Co.   Little Butte Volcanics.
      Refs. 161.
   5. Coal Creek, 18 mi. S of Oakridge, Lane Co.   Little Butte
      Volcanics.   Refs. 102, 161.

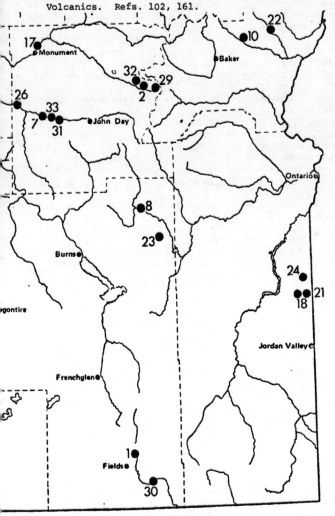

EARLY/MIDDLE MIOCENE FOSSIL PLANT LOCALITIES

6. Collawash River valley, Clackamas Co. ?Eagle Creek Fm.
   (Little Butte Volcanics). Collawash flora.
   Refs. 161, 242, 242a, 242c.
7. Dayville/Mt. Vernon Hwy., Grant Co. Mascall Fm. Mascall
   flora. Refs. 32i, 33.
8. Dry Creek, 12 mi. NE of Buchanan, Harney Co. Columbia
   River Basalt. Stinking Water flora. Refs. 33.
9. Eagle Creek valley, Hood River Co. Eagle Creek Fm. (Little
   Butte Volcanics). Eagle Creek flora. Refs. 32, 32a.
10. Keating, Baker Co. Columbia River Basalt. Sparta flora.
    Refs. 33, 67, 87.
11. Liberal, Clackamas Co. Little Butte Volcanics. Refs. 161,
    242a.
12. Little Butte Creek, 15 mi. NE of Medford, Jackson Co. Little
    Butte Volcanics. Refs. 161.
13. McCord Creek Bridge, Multnomah Co. Eagle Creek Fm. (Little
    Butte Volcanics). Eagle Creek flora. Refs. 32, 32a.
14. Maupin, Wasco Co. John Day Fm. Refs. 242a.
15. Moffatt Creek valley, Multnomah Co. Eagle Creek Fm. (Little
    Butte Volcanics). Refs. 32, 32a.
16. Molalla, Clackamas Co. Molalla Fm. (Little Butte Volcanics).
    Molalla flora. Refs. 12f, 53, 161, 242a, 242c.
17. Monument (3 mi. NE of), Grant Co. John Day Fm. Refs. 60, 242a
18. Rockville (near post office), Malheur Co. Sucker Creek Fm.
    Rockville flora. Refs. 191, 242c.
19. Ruckels Creek Bridge, Hood River Co. Sucker Creek Fm.
    (Little Butte Volcanics). Eagle Creek flora. Refs. 32, 32a.
20. Sandstone Creek valley, Clackamas Co. Little Butte Volcanics.
    Refs. 161, 242a.
21. Sheridan Ridge, 2 mi. E of Rockville, Malheur Co. Sucker
    Creek Fm. Rockville flora. Refs. 191.
22. Sparta, (5 mi. NW of), Baker Co. Columbia River Basalt.
    Sparta flora. Refs. 67, 87.
23. Stinking Water Mountains and Stinking Water Creek, Harney Co.
    Columbia River Basalt. Stinking Water flora. Refs. 33.
24. Succor Creek valley, Malheur Co. Sucker Creek Fm. Sucker
    Creek flora. Refs. 8, 27, 33, 68, 68a.
25. Tanner Creek valley, Multnomah Co. Eagle Creek Fm. (Little
    Butte Volcanics). Eagle Creek flora. Refs. 32, 32a.
26. Thomas Condon-John Day Fossil Beds State Park, Grant Co.
    (Mascall Ranch; Rattlesnake and Cottonwood creeks)
    Mascall Fm. Mascall flora. Refs. 27, 32c, 32d, 32i, 33.
27. Three Lynx Power Station, on Clackamas River, Clackamas Co.
    Little Butte Volcanics. Refs. 161, 242a.
28. Timbered Rock, 30 mi. N of Medford, Douglas Co. Little
    Butte Volcanics. Refs. 161.
29. Tipton (in railroad cut), Grant Co. Columbia River Basalt.
    Blue Mountain flora. Refs. 27, 33, 151.
30. Trout Creek, Harney Co. Trout Creek Fm. Trout Creek flora.
    Refs. 8, 33, 68, 68a, 126.
31. Van Horn Ranch, 12 mi. E of Dayville, Grant Co. (became
    Riverbank). Mascall Fm. Mascall flora. Refs. 32c, 32i,
    33, 103a, 111.
32. Vinegar Creek valley, Grant Co. Columbia River Basalt.
    Blue Mountain flora. Refs. 33.
33. White Hills, 9 mi. E of Dayville, Grant Co. (formerly
    Belshaw Ranch). Mascall Fm. Mascall flora. Refs. 32c, 33.

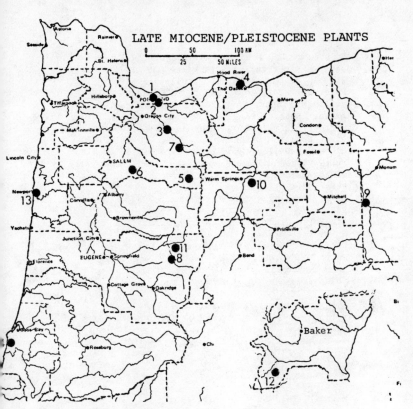

LATE MIOCENE/PLEISTOCENE PLANTS

LATE MIOCENE FOSSIL PLANT LOCALITIES
1. Buck Creek valley, 5 mi. E of Troutdale, Multnomah Co.
    Troutdale Fm. Troutdale flora. Refs. 32g, 32i.
2. Camp Collins (on Sandy River), Multnomah Co. Troutdale
    Fm. Troutdale flora. Refs. 32f, 32i.
3. Gazadero, at Faraday Power Station, Clackamas Co. Sardine Fm.
    Refs. 161.
4. Chenoweth Creek valley, 3 mi. NW of The Dalles, Wasco Co.
    The Dalles Fm. The Dalles flora. Refs. 32f, 32i.
5. Devil's Creek, near Breitenbush Hot Springs, Marion Co.
    Sardine Fm. Refs. 161.
6. Drift Creek, 5 mi. NE of Stayton, Marion Co. Sardine Fm.
    Refs. 161.
7. Fish Creek valley, Clackamas Co. Sardine Fm. Refs. 161, 242a.
8. Hidden Lake, S of Cougar Reservoir, Lane Co. Sardine Fm.
    Refs. 12f, 161.
9. Thomas Condon-John Day Fossil Beds State Park, Grant Co.
    (between Rattlesnake and Cottonwood Creeks). Rattlesnake
    Fm. Rattlesnake flora. Refs. 32d, 32i.
10. Vanora Grade, N of Madras, Jefferson Co. Deschutes Fm.
    Deschutes flora. Refs. 32e, 32i.
11. White Rock Creek, 5 mi. E of Cougar Reservoir, Lane Co.
    Sardine Fm. Refs. 161.
12. Unity, Malheur Co. "Ironside Fm". Refs. 113a.

PLEISTOCENE FOSSIL PLANT LOCALITIES
13. Newport, Lincoln Co. Coquille Fm. Refs. 12b, 12d.
14. Whiskey Run, S of Cape Arago, Coos Co. Coquille Fm.
    Refs. 12, 12b.

OREGON FOSSIL POLLEN

Despite its obvious utility in interpreting paleoenvironments, pollen has received little attention in the published literature of Oregon paleontology. Chaney (1959) notes that the wide range of habitats represented by a single pollen sample is a mixed blessing. Although pollen affords a wider view of paleoenvironments, care must be taken when attempting to unravel the story pollen tells with respect to the local environment.

Some of the most detailed records of floral change come to us in the form of prehistoric "floral successions" as recorded in pollen profiles. A "succession" is the gradual change plant communities undergo with time as a raw habitat develops through various phases up through a forest stage. The initial condition or pioneer phase is usually represented by grasses. The pioneer phase, after a period of dominance, prepares the soil to the degree that it is eventually displaced by an intermediate phase, often pine/fir forests. This latter flora, after an additional period of dominance and soil development, is replaced by the final or "climax community", frequently a hardwood or deciduous forest such as oak. Often floral successions are interrupted, slowed or even reversed by natural catastrophes such as climate changes, fires, floods, plant disease or insects. Since successions may develop from pioneer to climax phase in only a few hundred years, their recognition in the fossil record requires close interval soil samples and abundant fossils. Both conditions are rarely met with plant fossils other than pollen.

Because pollen and spores only occur very rarely in association with the parent plant, they are treated and classified as separate entities in most cases. We may, for example, be able to identify an individual pollen grain as being from oak (*Quercus*), but the association of a particular oak leaf species with oak pollen species usually remains beyond our grasp.

Often pollen for a given genus is far more distinctive than the leaf. The presence of that pollen in a fossil assemblage will, then, confirm the identification of a questionable leaf. Unfortunately described leaf and pollen floras from the same localities are rare in the literature. There are no published records of Mesozoic or Paleocene pollen in the State. One of the oldest pollen floras is that published by Hopkins (1967) wherein he describes material from the late Eocene Coaledo, the upper Eocene/lower Oligocene

Bastendorff, and the early Oligocene Tunnel Point Formations where they are exposed near Coos Bay, Oregon. The 34 "natural genera" Hopkins identified from these units suggest a warm, humid temperate climate here in the lower Tertiary. Rainfall is estimated at 50-60 inches per year with an annual temperature that rarely fell below freezing (32 °F.).

Middle Miocene microfloras are known from the Sucker Creek and Trout Creek Formations of eastern Oregon as described by Graham (1965). These pollen floras reflect the same climate and environment as the leaf floras from the same localities but show a much wider influence because of the nature of the combined pollen fossil record. The Sucker Creek and Trout Creek floras are indicative of upland lake environments with mild winter temperatures and rainfall of as much as 50-60 inches/year. Taggart and Cross (1974) produced pollen profiles from Sucker Creek which show a sudden shift in vegetation here from a mesic or wet to a xeric or dry environment.

Pollen from the Mascall Formation and the Columbia River Group are listed by Chaney (1959). In his study, use has been made of pollen as well as megafossils. The pollen reflects a modern forest dominated by trees which shed their leaves in winter. This includes angiosperms as well as the most numerous gymnosperms, *Taxodium* (bald cypress) and *Metasequoia* (dawn redwood).

Wolfe (1960, 1962) has described three separate Miocene microfloras from several localities in the northern Willamette Valley and Western Cascades. These three floras, the Collawash, the Molalla, and the Eagle Creek, are regarded by Wolfe as age equivalents and are all from the middle Miocene Little Butte Volcanic Series. In spite of their similarity in age, Wolfe interprets the Molalla flora as subtropical, and the Collawash and Eagle Creek as progressively more warm temperate. Frosts during this interval at these localities were not severe and rainfall was at 50-60 inches annually, similar to what it is today. Of particular interest in Wolfe's study are his speculations on the origin or source of individual floras. His interpretation of the floral changes between the Paleogene and Neogene in the Pacific Northwest focuses upon a major floral migration during this interval from the Cordilleran area westward. The stimulus for this migration may have been the progressive uplift of the Rocky Mountain area. The coastal Molalla flora with a subtropical signature received very few immigrants from the East during this floral migration, but the more warm temperate Collawash and Eagle Creek floras bear many migrant

FOSSIL PRODUCTS

pollen

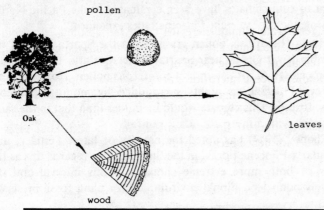

Oak

wood

leaves

FLORAL SUCCESSION

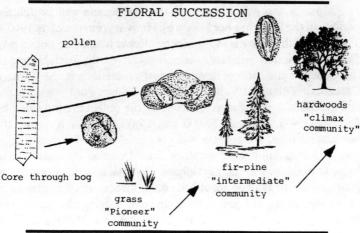

pollen

Core through bog

grass
"Pioneer"
community

fir-pine
"intermediate"
community

hardwoods
"climax
community"

LEAVES AND POLLEN

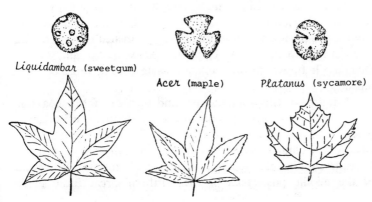

Liquidambar (sweetgum)

Acer (maple)

Platanus (sycamore)

73

species. These migrations have been traced through various floras across the Great Basin showing that with time many of the plants intolerant to cool climates have been eliminated in the Pacific Northwest while the cool tolerant Pinaceae types expanded.

Hoxie (1965) lists pollen grains from the Sparta flora of the Columbia River Group near Sparta, Baker County. This is an upper Miocene flora with abundant coniferous pollen. Hoxie suggests a valley of deciduous plants surrounded by mountainous coniferous forests. This climate would be milder than that in the same area today with a more abundant rainfall.

Chaney (1944) has noted the paucity of latest Tertiary, and particularly Pliocene floras, in the State, and he regards this as the effects of both more extreme climates in this interval and the relatively poor depositional environment for plant fossil preservation.

Most of the pollen work on the Pleistocene and postglacial periods in the State has been done by Hansen from the early 1940's. The bulk of this work is in Pleistocene floras where bog pollen profiles in association with large vertebrates reflect the glacial stages of this epoch in the form of several forest successions in the Western Cascades, Willamette Valley, and Coast Range. Four phases in the late glacial and postglacial reflected by the pollen are indicated by Hansen (1947). Between 20,000 and 15,000 years ago during the late glacial time, the climate was cool and moist. Lodgepole pine was at a maximum with hemlock, and Douglas fir expanded to become dominant in the next phase. Postglacial pollen profiles in coastal areas of Oregon record the dune succession from grasses to fir as the dunes are gradually stabilized. During the postglacial beginning warm, dry interval, Period II, between 11,000 and 8,000 years ago, the Douglas fir forests of the Willamette Valley and Coast Range expanded to a maximum then declined slightly. In the succeeding Period III of maximum dry and warmer temperatures, Douglas fir declined even more as oak flourished to a maximum. Then in the last Period IV, between 4,000 years ago and today, a Douglas fir forest returned as the climate became more cool and moist.

Within the Cascades the ash and pumice of Mt. Mazama, about 7,500 years ago, promoted the development of lodgepole pine forests that dominate the southern Cascades of the State today. As the climate became more warm and dry elsewhere in the Cascades, yellow pine replaced the lodgepole forests. Dry periods of the middle postglacial of the northern Great Basin in the

74

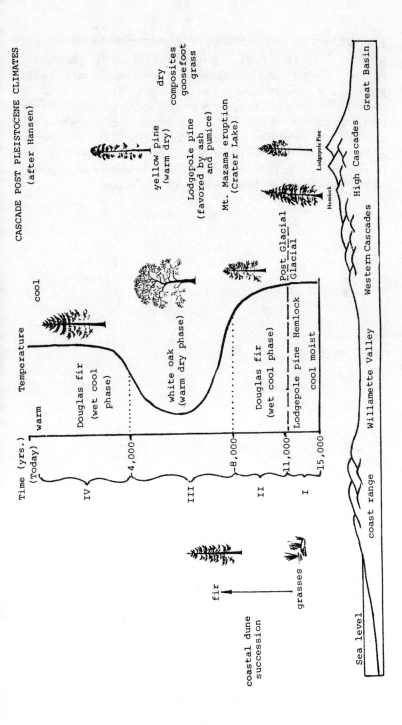

CASCADE POST PLEISTOCENE CLIMATES
(after Hansen)

Temperature

cool

warm

Douglas fir
(wet cool
phase)

white oak
(warm dry phase)

Douglas fir
(wet cool phase)

Lodgepole pine Hemlock
cool moist

Post Glacial
Glacial

yellow pine
(warm dry)

dry
composites
goosefoot
grass

Lodgepole pine
(favored by ash
and pumice)

Mt. Mazama eruption
(Crater Lake)

Hemlock

Lodgepole Pine

Time (yrs.)
(Today)

IV

4,000

III

8,000

II

11,000

I

15,000

fir

grasses

coastal dune
succession

Sea level

coast range

Willamette Valley

Western Cascades

High Cascades

Great Basin

75

Klamath area are nicely reflected in pollen profiles by the influx of grasses (Gramineae), goosefoot family members (Chenopodiaceae), and composites (Compositae). In many pollen profiles, fire appears to have had a dramatic effect on floras. In southern Washington State, for example, hemlock expansion in the postglacial period is apparently retarded by fire. Elsewhere fire promotes the expansion of certain forest types. For example, in eastern Oregon and northeast Washington state, white pine is favored by fire. In the southern Cascades, lodgepole pine favored by fire persisted over yellow pine in the same area.

OREGON FOSSIL INVERTEBRATES

Historically Oregon has long been known for its superb assemblages of fossil molluscs. The best preserved and most abundant assemblages are of Tertiary Age, but several diverse faunas occur in Mesozoic rocks here as well as in a few representative Paleozoic types. The Tertiary molluscan record is found west of the Cascades since post Mesozoic seas were restricted to that area. The geologic record of Tertiary seas in Oregon is one of a withdrawing or regressive ocean. The Eocene, in the lower Tertiary, is most widespread, with the Oligocene somewhat less represented. The Miocene, Pliocene, and Pleistocene are restricted to narrow strips in coastal exposures. Although Mesozoic rocks are not nearly as well represented as those of the Tertiary, they occur over a much wider area. Two principal areas of Mesozoic rocks are in eastern Oregon at Mitchell and Suplee/Izee and in the Klamath Mountains. Several paleogeographic maps showing the extent of Mesozoic seas have been drawn, but few take into account the displacement of the earth's tectonic plates. The lack of extensive exposures of Paleozoic rocks in Oregon precludes the precise location of old shore lines.

Before dealing with specific formations and fossil assemblages it is well to become familiar with the general morphology of the Tertiary and Mesozoic molluscs. The mechanical complexities of the clam shell with its ability to open and close have affected the morphology of the shell itself such that it is possible to make taxonomic distinctions with confidence on the basis of the shell. Snails with the single shell carried dorsally are not as clearly distinguished as to their taxonomic affinities from the shell alone.

Geologists working in Europe in the 19th century quickly saw the value of fossils as geologic time indicators. Early efforts to recognize European molluscan chronologies elsewhere in the world were limited in success because, like modern plants and animals, fossil organisms often tend to be provincial in their distribution. Very early, paleontologists realized that it was necessary to establish a local chronology of fossils to use in the immediate vicinity or basin then later to make correlations between various local sections to synthesize a master chronology. Molluscs are a group which are particularly provincial. This provinciality occurs despite the fact that many species of molluscs spend the earliest larval period of their lifespan as plankton free to be carried by ocean currents. Charles Lyell in 1833 recognized the difficulty of making

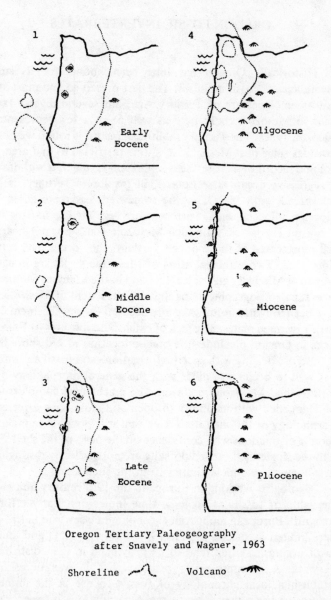

Oregon Tertiary Paleogeography
after Snavely and Wagner, 1963

Shoreline 〰〰〱   Volcano 🔥

correlations with provincial fossils and set about to employ the uniform faunal succession of the European Teriary. He defined the "Eocene," "Miocene" and "Pliocene" epochs on the basis of the number of fossil species in the rock unit that are found living to-day. To these epochs were added by subdivision and supplement the Pleistocene, Oligocene and Paleocene. In spite of the shortcom-

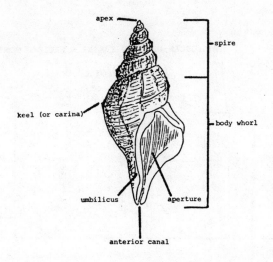

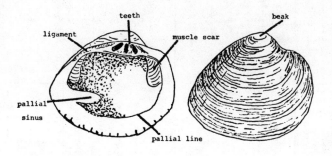

ings of Lyell's "percentage method" this strategy was used to initially recognize Tertiary epochs in California.

The West Coast has seen the establishment of several fossil chronologies because of the relative isolation of the province and the well developed fauna of every imaginable type. Early stage chronologies developed in California were referred to by the names of geologic rock formations bearing the characteristic fossils. The Vaqueros State is characterized, then, by the fossils that are found in part of the Vaqueros Formation. Unfortunately this Formation developed not only during the Miocene epoch but during the upper Oligocene epoch as well. The ensuing confusion here is between rock units or formations and time stratigraphic units or stages. A summary of the provincial stages used for the West Coast molluscan assemblages appears in Addicott (1976) along with a very recently established chronology for the Neogene.

MOLLUSCAN NEOGENE STAGES - PACIFIC NORTHWEST

(after Addicott, 1976)

TIME IN M.Y.	EUROPEAN STANDARDS		PROVINCIAL MOLLUSCAN STAGES		CALIFORNIA BENTHIC FORAMINIFERAL STAGES
			PACIFIC NORTHWEST	CALIFORNIA	
5	Neogene	Pliocene	Moclipsian	"San Joaquin and Etchegoin"	Venturian Repettian
					Delmontian
10		Miocene — Upper	Graysian	"Jacalitos"	Mohnian
			Wishkahan	"Margaritan"	
15		Miocene — Middle	Newportian	"Temblor"	Luisian and Relizian
20		Miocene — Lower	Pillarian	"Vaqueros"	Saucesian
			Juanian		
25	Paleogene	Oligocene — Upper	Matlockian (Armentrout, 1975)	Unnamed Addicott, 1973	Zemorrian

80

A final introductory aspect of this section is the use of molluscs other than as time indicators. Invertebrate fossils may be utilized in several ways to develop a prehistoric setting or paleoenvironment. If we examine a modern beach and continental shelf we see several changes in the marine environment as we proceed from the beach (strand) down across the shelf toward deeper water. The most obvious change is the gradual reduction of wave turbulance and the differences in the marine substrate, or bottom. Marine invertebrates respond to these as well as many other factors such as light penetration of the water, oxygen concentration, water temperature and turbidity. The tendency is for the invertebrate assemblages to change progressively toward deeper water. Fossil molluscs can then be used to extrapolate paleodepth as well as proximity to the strand. Some of these extrapolations can be carried out without referring to specific taxa. For example, we know that unless a clam is a burrowing type, those in the surf zone tend to have a thick, robust shell; whereas, those in deep quiet water are thin-shelled. Looking at entire faunas we are able to generalize that in the "specialized environments" of estuaries and bays where salinity and temperature may display extremes or may vary rapidly over short time intervals the diversity or number of species of invertebrates tends to be much smaller than in a "normal" marine environment. Diversity is also lower in the turbulent surf zone as well as on the outer continental shelf, in deeper water below wave base, and below the euphotic (light penetration) zone.

REEFS

Modern reefs of invertebrate accumulations usually include some frame building organism such as a coral and a sediment binding organism, frequently algae. In geologic terms a reef or "bioherm" is a very special submarine structure which implies that several environmental and biologic prerequisites have been met. Like a modern skyscraper, a reef must have a foundation or substrate to build upon. Organisms that construct this foundation are different from those that eventually make up the bulk of the overlying reef. In other words, something of a succession is implied. After building upon this substrate, the true bioherm must then rise by growth to the zone of wave action and maintain itself in that high energy zone. Again the corals that stud the rough windward side of the reef are altogether different from those lower down on the reef flanks.

The tendency in the geologic literature in the past has been to call rich accumulations of fossils "reefs." More often than not what is called a bioherm (reef) is in fact a "biostrome" or fossil rich lense without form or internal structure. Thick accumulations of invertebrates such as the clam, *Buchia*, in the southwest Oregon Jurassic/Cretaceous are old banks and not reefs. Similarly Lupher and Packard (1930) refer to pelecypod biostromes of Jurassic age in Grant County as reefs. Smith (1912) refers to coral reefs in the Triassic Eagle Creek area of Baker County in what are actually structureless accumulations of corals, pelecypods and cephalopods. Modern reefs are restricted to waters warmer than 58 °F and at a latitude lower than 20 °N and S. In western North America the lateral shifting of large scale tectonic plates must be taken into consideration before attaching any significance to fossil reef distribution. Reefs are typically found in association with limestone accumulations because both lithologic types share several environmental characteristics. The presence of carbonates (limestones) is progressively diminished in the Oregon Paleozoic and Mesozoic. In the Oregon Tertiary only a few localities of thin Eocene bioclastic (broken shell) limestones are known in the central Coast Range. Although biostromes are not uncommon in the State, true bioherms or reefs are unknown.

PALEOZOIC

Paleozoic rocks in eastern Oregon are part of an inlier surrounded by Mesozoic and Cenozoic units. The sequence of fossil bearing paleozoics in the eastern part of the State in Crook and Harney Counties includes Devonian through Permian rocks. The Devonian and Mississippian here are represented by limestone and chert in thicknesses up to 1,000 feet. Two limited exposures of unnamed Devonian limestones represent the oldest sediments in the State. Recently Johnson and Klapper (1978) and Savage and Amundson (1979) have described brachiopods and conondonts from these exposures, redefining the chronostratigraphy. Although the precise biologic affinity of conodonts is unknown, they display cosmopolitan distribution typical of pelagic organisms that makes them optimum for world-wide correlations. The best evidence sug-

Belodella

Ozarkodina

Icriodus

.1"

gests that conodonts were parts of a small fish-like vertebrate.

The thick Mississippian carbonate exposure of the Coffee Creek Formation in eastern Oregon bears brachiopods and coelenterates, but these invertebrate faunas have never been described in detail. A recent paper on some foraminifera from this formation (Sada and Danner, 1973) concludes that the Mississippian here is of Chester (late Mississippian) age.

PALEOZOIC CORALS

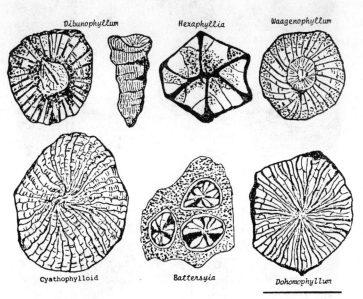

Dibunophyllum   *Hexaphyllia*   *Waagenophyllum*

Cyathophylloid   *Battersyia*   *Dohomophyllum*

Overlying the Coffee Creek Formation, the only apparent Pennsylvanian invertebrates that occur in the State consist of a very few specimens of shallow water clams, snails and crinoids in a marine facies of the Spotted Ridge Formation. The material was so

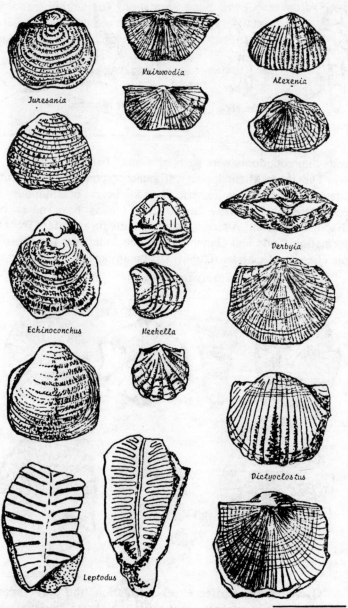

Juresania

Muirwoodia

Alexenia

Derbyia

Echinoconchus

Meekella

Dictyoclostus

Leptodus

limited and poorly preserved that an age designation was impossible (Mamay and Read, 1956).

Above the Spotted Ridge Formation a large fauna of Permian

brachiopods has been described in some detail by G. A. Cooper (1957) from the Coyote Butte Formation. That author was able to correlate Oregon faunas with middle Permian units elsewhere in North America. Cooper notes that Oregon Permian brachiopod

PALEOZOIC INVERTEBRATES

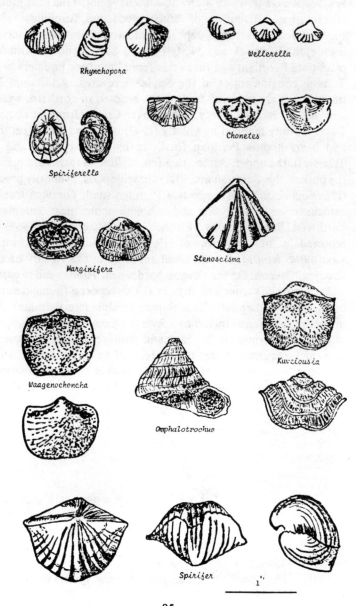

Rhynchopora

Wellerella

Spiriferella

Chonetes

Marginifera

Stenoscisma

Waagenochoncha

Kuvelousia

Omphalotrochus

Spirifer

1"

85

faunas show considerable affinity to faunas described from Russia. The distribution of invertebrates throughout the exposures of the Coyote Butte Formation suggests that the lower portion with abundant fusulinids, corals, and crinoids represents a more shallow paleoenvironment than the upper portion with abundant brachiopods. Elsewhere many authors have noted the near mutually exclusive distribution of brachiopods and fusulinids in late Paleozoic carbonate shelf deposits and have attributed it to increasing depth. Bostwick and Nestell (1965) have identified Guadelupean (late Permian) age fusulinids from limestone boulders in late Triassic conglomerates in the Suplee/Izee area. Additional Permian fossils are reported from rocks exposed in northern, western and eastern Baker County. The Clover Creek Group in northern Baker County mapped by Gilluly (1937) is around 4,000 feet thick and bears middle Permian fossils in its lower portion and late Triassic in the upper. Although a few molluscan fossils are present, the bulk of this Permian material is brachiopods, primarily productids. Yochelson (1961) reports a Permian snail, *Omphalotrochus*, associated with fusulinids and echinoid spines near Sumpter in northwest Baker County. Further east, Waterhouse (1968) has reported a new species of Permian productid brachiopod, *Kuvelousia leptosa*, from a small exposure in the vicinity of Cornucopia, Oregon. On the Oregon border near Homestead in eastern Baker County, Vallier and Brooks (1970) report a fauna identified by F. Stehli as Leonard-Guadelupian (middle late Permian). This invertebrate fauna including several groups of brachiopods, echinoderms (crinoids), bryozoa and molluscs (pelecypods) occurs in a thick sequence, 8,000-10,000 feet, of volcanic clastics, shales and limestones designated as the "Hunsaker Creek Formation." Locally within the unit the invertebrate fossils are abundant and well-preserved.

TRIASSIC

Although rocks of Triassic age are extensively exposed in western and central Oregon, Triassic fossils are well developed in only two areas in south central Wallowa County and in the Suplee/Izee area of southwest Grant County. Triassic faunas in the Wallowa Mountains are diverse and dominated by molluscs particularly cephalopods and pelecypods. Several species of corals are

86

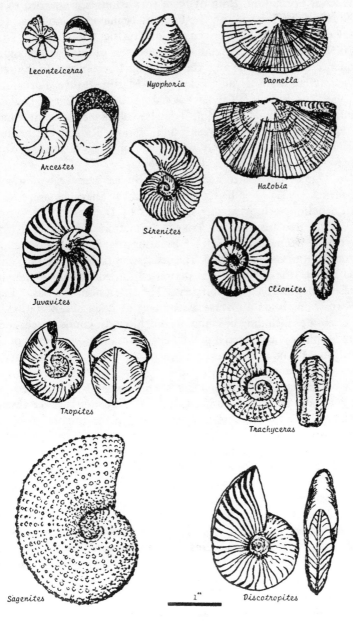

Leconteiceras

Myophoria

Daonella

Arcestes

Halobia

Sirenites

Juvavites

Clionites

Tropites

Trachyceras

Sagenites

1"

Discotropites

also reported along with a few brachiopods, crinoids, and sponges (Smith and Allen, 1941). Fossiliferous units here include the carbonate Martin Bridge Formation underlying the fine clastics of the Hurwal Formation. Both of these formations are assigned to the Norian (late Triassic) on the basis of cephalopod molluscs. In a review of the North American marine Triassic, Silberling and Tozer (1968) comment on the relative completeness of this late Triassic sequence in the Hurwal and Martin Bridge Formations. In the same vicinity stratigraphically below the Martin Bridge Formation, Vallier and Brooks (1970) have noted up to 15,000 feet of Ladinian and Karnian (middle and late Triassic) volcanic clastics informally designated the "Doyle Creek" and "Grassy Ridge" Formations. The latter age assignments were made on meagre shallow water pelecypods and cephalopods.

One of the earlier studies on the Suplee/Izee Triassic was by Schenck (1931) when he reported on local stratigraphy and fossils. In a definitive look at this same area by Dickinson and Vigrass (1965), three fossiliferous Triassic formations are reported from oldest to youngest: the Begg Formation, the Brisbois Formation, and the Rail Cabin Argillite. The occurrence of fossils in these units as well as the lithology has permitted an evaluation of the successive depositional environments. The oldest unit, the clastic Begg Formation, bears a diverse assemblage of open ocean marine invertebrates including belemnoids, brachiopods, corals, pelecypods, and cephalopods. Overlying the Begg Formation, limestones of the Brisbois Formation are even more fossiliferous with a marine invertebrate fauna typical of a warm shallow, upper neritic zone on a carbonate shelf. Oysters and frame building corals in some beds occur in near reef-like accumulations, but true bioherms are absent. The Rail Cabin Argillite (shales) overlying the Brisbois carries a somewhat more diminished fauna than the units below and is represented by pod-like occurrences of limestones within the argillite of what may have been banks of offshore carbonate knolls. These bear a fauna dominated by molluscs including cephalopods and pelecypods.

Triassic fossils have been reported from exposures of greywackes, shales and volcanic tuffaceous rock of the Aldrich Mountains Group in the area between Izee and John Day. Ammonites of the genera *Placerites*, *Sandlingites*, and *Vredenburgites* in the Murderers Creek Graywacke are Norian (upper Triassic). Overlying the latter unit is the Keller Creek shale bearing *Arnioceras*, *Crucilobiceras*, and *Gleviceras* of Sinemurian (lower Jurassic) Age (Brown and Thayer, 1966).

/

JURASSIC

Fossiliferous Jurassic marine rocks in Oregon are scattered in a southwest to northwest trending series of exposures across the State. The southwest most exposures occur on the coast and inland at Curry County in the Port Orford and Gold Beach areas and have been described in some detail by Koch (1966) as well as by other authors. Scattered microfossils including foraminifera and radiolaria are found in only very limited amounts in the Galice, Rogue, and Dothan complex here as well as in the Colebrooke. Ramp (1969) reports an occurrence of the Jurassic species *Buchia piochii* from questionable Dothan exposures on Boulder Creek near Chetco River, southern Curry County. The Galice has been dated as late Oxfordian to middle Kimmeridgian by the occurrence of ammonoids and by the pelecypod *Buchia concentrica* in inland exposures of the unit. The overlying Riddle Formation of the lower Myrtle Group in the Days Creek area has several fossiliferous horizons that permit an uppermost Jurassic Tithonian stage assignment. Fossil assemblages include a curious mixture of the shallow water, mud facies, oyster-like pelecypod *Buchia* as well as several open ocean indicators including radiolaria, ammonites, and belemnites. The lower Myrtle Group in the coastal area is represented by the Otter Point Formation of cherts, graded clastics and volcanics of over 10,000 feet in thickness. Locally the unit bears belemnoids and *Buchia piochii* which places it in the Tithonian (uppermost Jurassic). Several of these molluscs, in particular species of *Buchia* and the ammonites, are correlative with established zones in California (Imlay, et al., 1959).

Further east in Douglas County, fossiliferous sections of the Riddle Formation have been described (Imlay and Jones, 1970; Imlay, et al., 1959) where it crops out south of Roseburg in the valley of the South Umpqua River. Originally described by Diller (1907), the clastics of the Riddle Formation bear fossil specimens locally of rich accumulations of the clam *Buchia*, referred to as "*Aucella*" by Diller, as well as scattered ammonites including *Proniceras*, *Spiticeras*, and *Protocanthodiscus*.

Jurassic exposures in the Suplee/Izee area include several members and formations dominated by marine clastics and volcanigenic sediments interspersed with thin limestones representing a wide range of environments. Described by Lupher (1941) and later in more detail by Dickinson and Vigrass (1965), the Jurassic

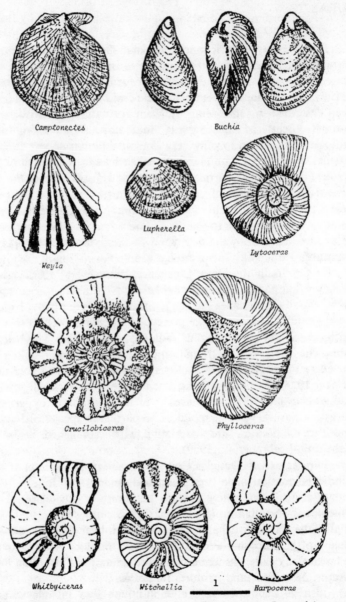

Camptonectes

Buchia

Weyla

Lupherella

Lytoceras

Crucilobiceras

Phylloceras

Whitbyiceras

Witchellia

1

Harpoceras

here ranges from the Hettangian (lowermost Jurassic) stage through the Callovian (uppermost Jurassic) stage. Five major lithologic units comprise the Jurassic including from oldest to youngest, the Greylock Formation, the Mowich Group, the

Snowshoe Formation, the Trowbridge Formation and the Lonesome Formation. Zonal ammonites are summarized by Imlay (1973). The brachiopod, *Discina*, in the Greylock Formation suggests brackish, marine conditions and possible estuarine environments. In addition to the brachiopods, several genera of ammonites are also known from this formation: *Paracaloceras*, *Phylloceras*, *Psiloceras*, and *Waehneroceras*.

The Mowich Group includes four formations which become less calcareous and more coarsely clastic upward. The lowermost, Robertson Formation, including biostromal limestone interbedded with volcanic sandstone, is characterized by accumulations of the elongate pelecypod, *Plicatostylus*, often in growth position. Accompanying these are a diverse assemblage of other pelecypods including *Trigonia*, *Pholadomya*, *Lima*, *Astarte*, and *Lucina*. Along with this very shallow open shelf assemblage are found scattered ammonites of the genus *Weyla*. Overlying the Robertson, the calcareous, sandy Suplee Formation is also characterized by a shallow shelf assemblage of molluscs including several of the same forms from the Robertson Formation as well as the common Jurassic, oyster-like form, *Gryphaea*. Ammonites are rare in the Suplee, but in the overlying Nicely Formation they occur in abundance permitting the correlation of that outer shelf unit with the Toarcian stage of Europe. In contrast to the underlying Jurassic units, the Hyde Formation is relatively unproductive of fossils except for some poorly preserved ammonites and brachiopods which may represent deep water. Dickinson and Vigrass note the progressive deepening paleoenvironments of the Mowich Group and regard this to be the result of a transgressive sea. Beginning with a shelf carbonate environment followed by outer shelf clastics with pelecypod assemblages, the succession ends with offshore open ocean ammonite faunas in the upper part.

Overlying the Mowich Group, the four members of the Snowshoe Formation are primarily clastics with some thin limestones. Dickinson and Vigrass have interpreted the environment of the Snowshoe Formation as representing a shallow marine platform. Locally abundant ammonites place the Formation in the Bajocian stage (early middle Jurassic). Pelecypods are present throughout the Formation, and, although they represent locally shallow conditions, they do not accumulate in biostromal quantities as in the underlying Mowich Group.

The uppermost Jurassic units in the Suplee/Izee area are the Trowbridge and Lonesome Formations. Unlike the Mowich Group

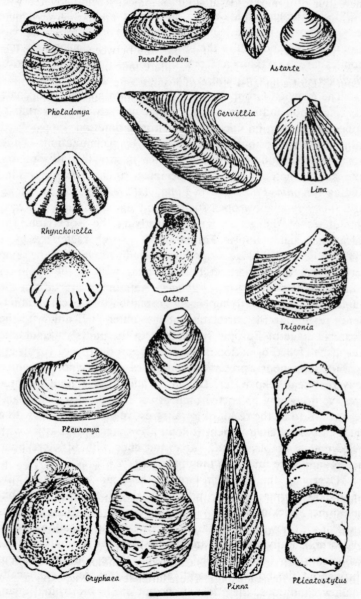

Parallelodon

Astarte

Pholadomya

Gervillia

Lima

Rhynchonella

Ostrea

Trigonia

Pleuromya

Gryphaea

Pinna

Plicatostylus

and the Snowshoe Formation, the Trowbridge and Lonesome Formations bear only limited fossil assemblages of scattered pelecypods and ammonites of Callovian (uppermost middle

Jurassic) stage. Dickinson and Vigrass interpret these formations as indicative of a marine basin floor periodically being covered with "turbidites", the periodic downslope rapid submarine movement of sand and mud mixtures.

In northern Harney County halfway between the Suplee/Izee area and Burns, less than half a square mile of exposure of we!'-indurated ironstained fossiliferous sandstone with lesser amounts of shale and conglomerate comprises about 2,500 feet of section and are designated the Jurassic Donovan Formation by Lupher (1941). The Donovan bears endoceratid and echioceratid late Sinemurian (lower Jurassic) ammonites, *Uptonia*, *Coeloceras*, *Metechioceras*, and *Deroceras* along with an upper neritic shallow marine pelecypod assemblage inluding *Plicatostylus*, *Pecten*, *Pholodomya*, *Pleuromya*, *Pinna*, *Modiolus*, and *Gervillia*. The striking, well-oxidized fossiliferous sandstones of the Donovan Formation have been confused previously with a similar facies of the Hardgrave sandstone in the Taylorsville region of California.

Further east of the Suplee/Izee area, Jurassic fossiliferous exposures are scattered in northern Malheur County, in southern Baker County, and in Wallowa County. Localities in the Clover Creek and Juniper Mountain areas of Malheur County as well as in the Huntington area of Baker County have yielded several middle Jurassic ammonites including the genera *Tmetoceras*, *Witchellia*, and *Stephanoceras* (Imlay, 1973; Wagner, et al., 1953). Exposures of clastics mapped as Triassic Hurwal Formation by Smith and Allen (1941) also yield Pliesbachian (middle early Jurassic) stage pelecypods of the genus *Lupherella* (Imlay, 1967). That author interprets the genus as a shallow dwelling organism on mud bottoms. These exposures are correlative with the Jurassic exposures within the Suplee/Izee area and represent a more clastic portion of the early Jurassic shoreline than is characterized at Suplee/Izee by biostromal accumulations of the clam, *Plicatostylus*. A brief mention by Morrison (1964) of upper Jurassic mudstone in the Snake River Canyon in Wallowa County represents the northeastern most exposure of Jurassic in the State. Black mudstones here of the Coon Hollow Formation have poorly preserved specimens of ammonites identified by Imlay as *Carcinoceras* indicative of lower Oxfordian (late Jurassic) stage. These latter few Jurassic exposures to the east and northeast of the excellent Suplee/Izee exposures bear neither prolific or well-preserved fossils. What they do show is the continuous trend of the Jurassic shoreline throughout the period as it swings northeastward from Suplee.

CRETACEOUS

Like the underlying Jurassic, the Cretaceous units in the State are scattered in a southwest-northeast trend across Oregon. The eastern limits are, however, not as extensive as the Jurassic. Exposures of the lower Cretaceous Myrtle Group in the Port Orford-Gold Beach areas of northern Curry County include the Rocky Point Formation and the Humbug Mountain Conglomerate (Koch, 1966). Fossils occur scattered throughout the units and are characterized by faunas of the shallow water pelecypod, *Buchia*, displaying beach abrasion. Belemnites (*Aulacoteuthis*), ammonoids (*Sarasinella*, *Kilianella*), and the pelecypod species, *Buchia crassicolis*, have been used to assign these units to the Valanginian (lower Cretaceous) stage. Open ocean forms along with sedimentary structures suggest that although some intervals of the Myrtle Group were deposited in shallow water, they were later displaced into deep water by submarine mudslides and turbidity flows.

Fine and course clastics of the Days Creek Formation of the Myrtle Group crop out in a broad area south of Roseburg along the Umpqua tributaries and in southern Jackson County. Imlay (et al., 1959) has employed ammonites and the pelecypod genus, *Buchia*, to assign the Days Creek to the Valanginian through the lower Hauterivian stages of the lower Cretaceous. Locally in the Days Creek area the genus, *Buchia*, occurs in biostromal accumulations in what were clearly very shallow water environments. A thin Valanginian sandstone near Agness yields the pelecypod, *Buchia crassicollis*, as well as the ammonites, *Simbirksites*, *Hollisites*, and *Hoplocrioceras* (Popenoe, 1960). Jones (1969) has succeeded in zoning the Myrtle Group (upper Jurassic and lower Cretaceous) by characteristic species of the genus, *Buchia*.

To the southeast of Roseburg in Jackson County, scattered Cretaceous exposures of the Hornbrook Formation described by Peck (et al., 1956) include well indurated, well bedded clastics of late Cretaceous age. Considerable time separates these relatively young Cretaceous units from the lower Cretaceous units to the west and north in the coastal area and in the Umpqua Valley area south of Roseburg. In drawing paleogeographic maps it is well to bear in mind that the Hornbrook Formation represents a much younger sea than that of the Days Creek area.

The Hornbrook Formation consists of coarse clastics in its lower half with shallow water molluscs including the snail, *Tur-*

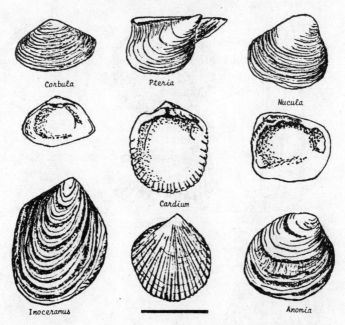

Corbula   *Pteria*

Nucula

Cardium

Inoceramus   *Anomia*

ritella, and the pelecypods, *Inoceramus* and *Trigonia*, in biostromal quantities. The upper half of the formation is primarily fine clastics with open ocean nektonic ammonoids. The latter fossils have been used by Peck (et al., 1956) and Popenoe (et al., 1960) to assign the Hornbrook Formation to a broad interval from the Cenomanian through the Campanian stages of the upper Cretaceous. In a recent paper, Ward and Westerman (1977) report the first occurrence in America of the ammonite genus, *Nipponites*, from the Hornbrook Turonian stage sediments. This peculiar un-coiled species of cephalopod was apparently neutral to barely buoyant and existed as a planktonic or floating form. The upper-most Cretaceous exposures in the State are near the coast in Curry County on the Pistol River where Popenoe (et al., 1960) notes the occurrence of late Campanian/early Maastrichtian stage molluscs including *Inoceramus, Meekia, Cymbophora* and *Mytilus*.

Eastern Oregon Cretaceous exposures are scattered over broad areas in Wheeler, Crook and Grant Counties. Jones (1960) has shown that the Albian (upper lower Cretaceous) stage exposures in the Mitchell area are correlative with an exposure on Grave Creek in southwestern Oregon in Jackson County permitting a paleogeographic reconstruction of the Albian shoreline. Wilkinson

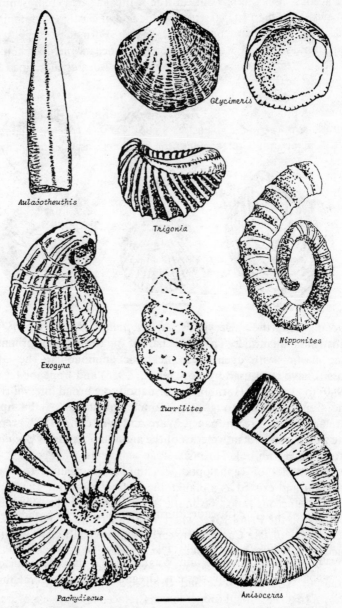

Aulacotheuthis

Glycimeris

Trigonia

Exogyra

Turrilites

Nipponites

Pachydiscus

Anisoceras

and Oles (1968) have described the Cretaceous rocks of the Mitchell area as a marine mudstone, the Hudspeth Formation, interfingering in the northeast with the fluvial/deltaic coarse clastics of the

Gable Creek Formation. More than seventy square miles of Cretaceous rocks crop out in the Mitchell area representing somewhat less than 10,000 stratigraphic feet of clastics. Ammonites and other molluscs from the Hudspeth Formation suggest a shallow mud shelf environment open to the ocean. Included here is a new species of the uncoiled ammonite, *Anisoceras*, described by Packard and Jones (1965).

Cenomanian (lowermost upper Cretaceous) fossiliferous sediments designated as the Bernard Formation by Dickinson and Vigrass (1965) crop out in eastern Crook County. These sediments represent the only Cretaceous in the Suplee/Izee area and include some 1,500 feet of sandstones and conglomerates fossiliferous in the lower third. Several ammonite species in the invertebrate fauna identified by R. W. Imlay and D. L. Jones as *Turrilites*, *Desmoceros*, and *Anthonya* place the fossiliferous interval in the lower Cenomanian. Nineteen species of invertebrates from the Bernard Formation are listed including several shallow water pelecypods, *Trigonia*, *Exogyra*, *Inoceramus*, and *Ostrea*, as well as other molluscs and echinoderms. Thick-shelled pelecypods such as *Trigonia* and the oyster-like *Exogyra* and *Ostrea* suggest a wave washed, shallow, upper neritic marine environment.

The eastern most exposures of marine Cretaceous in Oregon were noted by Mobley (1956) when he described beach worn fossils in the Dixie Mountain vicinity of eastern Grant County. This meagre fauna includes *Meekia*, *Anomia*, a naticoid snail, and *Aphrodina* (*Meretrix*), dated as Turonian by J. B. Reeside of the U. S. National Museum.

PALEOGENE

Tertiary marine rocks are confined to a steadily decreasing shelf area west of the Cascades during the period. The broad shelf areas of the Eocene and Oligocene largely disappear with the Miocene volcanic activity and late emplacement of the high Cascade volcanics. Snavely and Wagner (1963) have constructed paleogeographic maps of western Oregon for this interval that show this sea withdrawal.

Paleocene rocks have been identified in the State from foraminiferal faunas, but no megafossils of unquestionable Paleocene age have been reported.

Eocene fossiliferous marine rocks are widely scattered in western Oregon, but four areas receiving specific attention include Coos Bay, Douglas County, Polk County and the Nehalem River area west of Portland. The oldest Eocene exposures in the State are to be found in the Roseburg area and in the Umpqua River valley. Baldwin (1974) has subdivided the old Umpqua Formation of Diller (1898) to include in ascending order: the Roseburg, Lookingglass and Flournoy Formations. Megafossils are rare in the Roseburg and Flournoy Formations, but a small mid-neritic Flournoy mollusc fauna found in association with decapods was listed by Orr and Kooser (1971) and included *Acila*, *Anomia*, *Crassatella*, *Glycimeris*, *Nuculana*, *Ostrea*, *Solena*, *Tellina*, *Fusinus*, *Homalopoma*, *Mitra*, *Olivella*, *Siphonalia*, *Turritella*, and *Volutocorbis*. Lookingglass faunas of molluscs have been described by Turner (1938) in some detail from near Glide at the mouth of Little River in what is probably the best Eocene invertebrate locality in Oregon. The well preserved, diverse tropical assemblage of molluscs includes primarily pelecypods and gastropods and is indicative of very shallow water with a sandy substrate. A characteristic fossil at this locality is the large thick-shelled clam, *Venericardia hornii*. Turner (1938) was able to correlate the old Umpqua here with rocks in California of the Capay (Domengine) stage.

The Tyee Formation uncomformably overlying the Flournoy is usually unfossiliferous, but a fauna of some 20 species has been described by Turner (1938) and revised by Hoover (1963).

Coos Bay Eocene mollusc bearing rocks, primarily the Coaledo and lower Bastendorff Formations, have been described by several authors. Turner (1938) examined faunas of both upper and lower Coaledo and correlated it with the Tejon (Narizian Stage) of California. Weaver (1945) in a summary of Coos Bay Tertiary formations extrapolated a temperature range of 60 °F to 80 °F on the basis of certain molluscan genera. The Coos Bay area was mapped in detail (Allen and Baldwin, 1944; Baldwin, 1973) to evaluate the coal resources there including a review of the stratigraphy and paleontology of local formations. Blake and Allison (1970) briefly treated ophiuroids (brittle stars) from the Coaledo Formation, but the most definitive paleoenvironmental picture for the area was developed by Dott (1966). In his assessment of the Coaledo Eocene deltaic sedimentation that developed the coal, Dott notes the presence of several molluscan genera that characterize subtidal and intertidal zones as well as being common

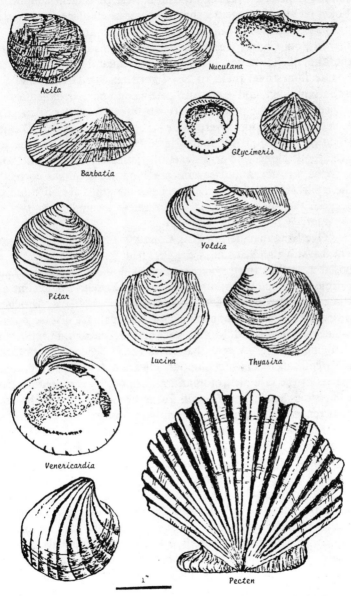

Acila

Nuculana

Barbatia

Glycimeris

Yoldia

Pitar

Lucina

Thyasira

Venericardia

Pecten

1"

to brackish bays including *Conus, Crepidula, Pachydesma, Polinices* and *Solen*.

Coaledo equivalent rocks of the Spencer Formation in Lane

County are mentioned by Turner (1938) in his review of Oregon Eocene molluscs. Spencer exposures west of Eugene mapped by Vokes (et al., 1951) yield a shallow water assemblage characterized by the genera *Acila*, *Tellina*, *Pitar*, *Turritella* and *Polinices*.

In Polk County west of Salem, exposures of Eocene rocks of the Yamhill Formation include the only quantity of Tertiary carbonate (limestone) rocks of any significance in Oregon and bear importantly on mid Cenozoic paleogeography. Sedimentary interbeds within the pillow basalts and breccias of the Siletz River Volcanics yield an invertebrate fauna identified by Durham (Baldwin, 1964). This fauna includes several gastropods and pelecypods along with fragments of echinoderms and coral, some bryozoa, crustacea and several brachiopods. Durham correlated these Siletz elements with the Eocene Capay stage of California and suggests that the depositional environment was in less than 60 feet of water.

Overlying the Siletz in Polk County is the Yamhill Formation. The basal Yamhill includes the Rickreall Limestone Member exposed at Dallas, Buell Quarries, and at Boulder Camp and dated variously by molluscs and foraminifera as Eocene Domengine and Ulatesian/Narizian Stages. Invertebrate populations associated with the bioclastic carbonates of the Rickreall are diverse but not usually well-preserved. Molluscs are best represented, but the fauna also includes crustacea (barnacles and decapods), bryozoa, calcareous algae and brachiopods. Hickman (1976) has reported a species of the rare pleurotomarid gastropod, *Pleurotomaria*, from a limestone lens in the Yamhill Formation and notes the habitual occurrence of this genus in proximity to submarine oceanic basalts here and elsewhere in Japan. Boggs, Orr, and Baldwin (1973) have interpreted the Rickreall Limestone occurrence within the Yamhill Formation as a series of offshore submerged carbonate banks in the zone of high wave energy.

Uncomformably overlying the Yamhill Formation and in places the Siletz River Volcanics, the Upper Eocene Nestucca Formation consists of varying thicknesses from 800 feet up to 7,500 feet of thin-bedded tuffaceous siltstones with intervals of sandstones, basalt flows and breccias. Although the unit has been characterized in part as "brackish" (Baldwin, 1976), the limited molluscan fauna it bears including *Acila*, *Nuculana*, *Yoldia*, *Euspira*, *Homalopoma*, *Cymatium*, *Conus* and *Exila* would indicate a shelf assemblage below the intertidal zone. Foraminifera from the Nestucca indicate an outer neritic to upper bathyl environ-

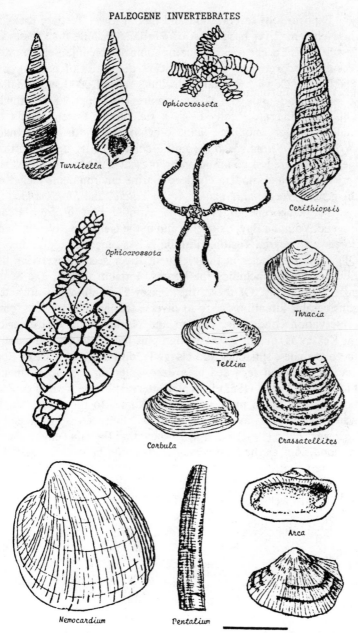

Turritella

Ophiocrossota

Cerithiopsis

Ophiocrossota

Thracia

Tellina

Corbula

Crassatellites

Nemocardium

Pentalium

Arca

ment. The Nestucca Formation has been correlated with the late Eocene Spencer Formation of the molluscan "Tejon" stage and with the upper Narizian foraminiferal stage.

101

Exposures of several thousand feet of lower Tertiary rocks in the Nehalem River basin west of Portland include thicknesses of fossiliferous Eocene rocks including some petroleum reservoir units recently discovered to bear significant quantities of natural gas. The Eocene Cowlitz Formation includes some 1,000 feet of shales, siltstones, and sands with a basal conglomerate up to 200 feet thick. This conglomerate locally bears a rich upper Eocene fauna of shallow water molluscs and brachiopods with the genera *Glycymeris*, *Ostrea*, *Mytilus*, and *Acmaea*. In its upper portion megafossils are less common and represent a mixture of brackish, very shallow and moderately deep marine environments including the genera *Nuculana*, *Nucula*, *Acila*, *Yoldia*, and *Siphonalia*.

The Oligocene of Oregon is exposed best in four areas: Eugene, Yaquina Bay, Coos Bay and in the Nehalem River basin of northwest Oregon. Tectonic instability has generated a diversity of Oligocene lithofacies in the Nehalem River section overlying the Eocene Cowlitz including the Keasey, Pittsburg Bluff, and Scappose Formations. Of these, the Keasey Formation has attracted considerable attention due to its diverse faunas and excellent fossil preservation conditions. Warren and Norbisrath (1946) describe the Keasey as roughly 2,000 stratigraphic feet of poorly bedded tuffaceous shales, siltstones and clays. Traditionally the Keasey Formation has been regarded as a deep-water marine environment. Moore and Vokes (1953) interpret intervals of the Keasey deposition to have taken place in close proximity to the shore, but in water well below wave base as deep as 3,000 feet. More recently, however, Zullo (1964) was able to show that the same environmental conditions as the Keasey represents could be met in water less than half that depth. One of the best known Keasey localities is near Mist, Oregon (Moore and Vokes, 1953), from which many particularly well-preserved specimens representing two species of the articulate Tertiary crinoid genus, *Isocrinus*, have been collected. A rich and well-preserved deep-water molluscan assemblage from the Keasey has been listed but not described in detail in the literature. Commoner genera include *Lima*, *Acila*, *Olequahia*, *Epitomium*, *Nemocardium*, *Bruclarkia*, *Solemya*, and *Pitar*. In addition to the crinoids and molluscs, Zullo (1964) notes in the Keasey local fauna at Mist an abundance of corals, *Flabellum*, fragments of ophiuroids (brittle stars), asteroids (starfish), decapods (crabs), fish scales, plant fragments and a new species of the echinoid genus, *Salenia* (*Salenia schencki*). Hickman (1972) has described a new species of Keasey gastropod of a rare genus, *Phanerolepida*.

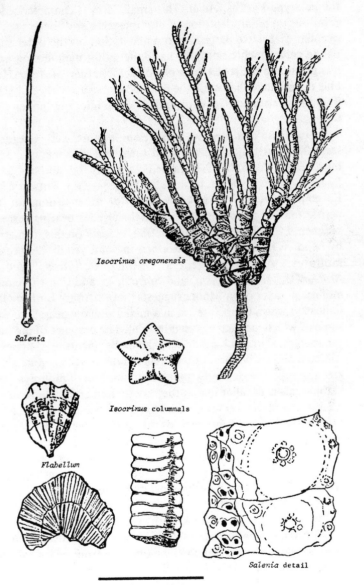

Salenia

Isocrinus oregonensis

Isocrinus columnals

Flabellum

Salenia detail

This genus is presently known only from a few deep-water environments in the Holocene and Tertiary of the western Pacific Basin, and Hickman notes that fossil deep-water molluscan assemblages are poorly known in general.

103

A common element in the Oligocene and Miocene interval is the pelecypod genus, *Acila*. This small form is characterized by a primitive taxodont dentition of multiple teeth and sockets and by a peculiar divaricate pattern of ornamentation on the outer surface of the valves. Schenck (1936) has done a thorough monograph on this genus and proposed three biozones for species of *Acila*. He was able to conclude that while the form occurs in a variety of habitats, it is most common in cool and cold middle and upper neritic waters.

Overlying the Keasey Formation in northwest Oregon are some 800 stratigraphic feet of sandstones and siltstones of the Pittsburg Bluff Formation. Regarded variously by authors as late Eocene/early Oligocene to middle Oligocene, the Pittsburg Bluff Formation bears a remarkable diversity of environments from neritic marine to intertidal and has local coal bearing terrestrial sediments. The most recent and definitive work on this formation is by Moore (1976) who has described in some detail 48 species of molluscs. Abundant genera include *Polinices*, *Neverita*, *Bruclarkia*, *Acila*, *Callista*, and *Spisula*. In addition to details of molluscan taxonomy, Moore suggests that Pittsburg Bluff molluscs indicate a warm temperate sea in a mixed shallow nearshore water, some of which may have been intertidal. The presence of molluscan genera typical of sand, mud flats, and embayments in its upper half and coal and carbonaceous sediments in the lower suggests a transgressing sea. The rich biostromal accumulation of molluscan shells in the sands is peculiar in that they are seldom articulated and show signs of transport, yet most bear little or no evidence of beach abrasion on the shells. Moore regards this as evidence of storm periods catastrophically sorting the shells up upon beaches from habitats as deep as 75-150 feet. In some of the fossiliferous slabs accumulated in this way up to 50% of the matrix may be calcium carbonate shell material. In addition to echinoids, fish remains are found in the Pittsburg Bluff including "otoliths", or earbones of conger eels and rat tails, as well as shark teeth representing seven genera.

Stratigraphically near the Keasey/Pittsburg Bluff boundary, exposures of the Gris Ranch Conglomerate from near Clatskanie, Oregon, and Deer Island near St. Helens bear a diverse shallow water assemblage completely unlike the Keasey and bearing only a few species in common with the Pittsburg Bluff (Moore, 1976). This fauna is apparently indicative of an intertidal paleoenvironment, and some of the more common genera include *Glycimeris*, *Arca*, *Loxocardium*, and *Puncturella*.

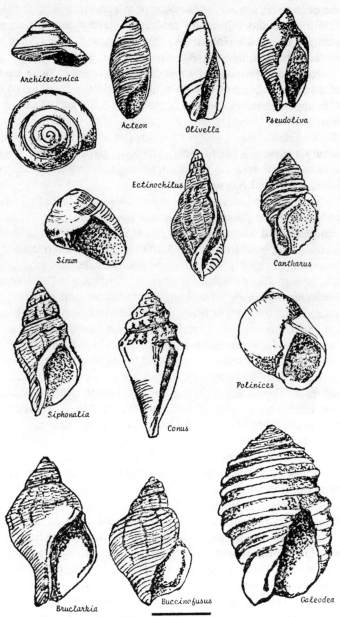

The Scappose Formation overlying the Pittsburg Bluff in northwestern Oregon consists of 1,500 feet of sandy shales of upper

Oligocene to lower Miocene Age. Only a single species of snail and no pelecypods co-occur in the Scappose and underlying Pittsburg Bluff Formations. Common molluscan genera in the Scappose listed (Warren and Norbisrath, 1946) include: *Acila*, *Nuculana*, *Yoldia*, *Anadara*, *Spisula*, *Bruclarkia*, *Pitar*, and *Macoma*.

Limited exposures of Oligocene invertebrate fossil bearing sediments occur in the northen Willamette Valley in scattered outcrops around Salem as well as in the Scotts Mills area (Peck, et al., 1964; Harper, 1946; Orr and Faulhaber, 1975). Referred to as the "Butte Creek Beds" where they crop out in northern Marion County, these localities bear shallow marine water mollusc dominated assemblages. Common genera are *Neverita*, *Bruclarkia*, *Solen*, *Pitar*, and *Nuculana*.

At the same latitude, outcrops of Oligocene rocks in Yaquina Bay include the Alsea and Yaquina Formations. The Alsea has been described (Snavely, et al., 1975) as fossiliferous marine siltstones and sandstones up to 1,100 feet thick. Concretions in the formation commonly bear fossil decapods (crabs) and molluscs. A rich foraminiferal fauna as well as molluscs of the genera *Bruclarkia*, *Acila*, *Parvicardium*, *Macrocallista*, and *Turritella* suggest a paleoenvironment beginning in the lower part of the Formation with bathyal to outershelf environments which become gradually shallower with time to near littoral conditions at the top. Cool water temperatures are reflected by the foraminifera as well as by a time duration from the Refugian through much of the Zemorrian stage. The Yaquina Formation in the same vicinity consists of over 1,500 feet of deltaic well-bedded and locally cross-bedded sandstones. Locally bearing foraminifera, abundant molluscs, fish scales and vertebrates as well as a cetacean (whale) remains, this unit is regarded as late Oligocene to early Miocene.

South of Yaquina Bay at Coos Bay, Oligocene units include the upper Bastendorff and Tunnel Point Formations. The Bastendorff consists of nearly 3,000 feet of shales and tuffaceous siltstones. Foraminifera are abundant and have been used by various authors to assign the formation to both the late Eocene and early Oligocene. Weaver (1945) regards the unit as wholly Eocene and reports very few molluscs except poorly preserved species of *Acila*. Foraminifera suggest intermediate to outer continental shelf deposits. The Tunnel Point Sandstone at Tunnel Point, Coos County, comprises some 800 feet of fine-grained tuffaceous sandstones comformably overlying the Bastendorff Formation. Weaver (1945) reports 33 species of molluscs scattered throughout the unit at its only known outcrop. The molluscan assemblage is dominated by the genera *Spisula*, *Molopophorus*, *Bruclarkia*, *Pachydesma*, and *Tellina* and clearly reflects a shallow upper shelf environment.

In the southern Willamette Valley exposures of lower middle Oligocene tuffaceous marine siltstones and sandstones of the Eugene Formation crop out in a broad area around Eugene, north along the Valley, and around Salem. Several authors have treated the molluscs of this formation, but the most definitive work to date is that of Hickman (1969). In describing the invertebrates of the Eugene Formation, she reported some 67 species of molluscs along with specimens of ophiuroids, echinoids, barnacles, foraminifera, decapods (crabs), brachiopods, and fish remains (shark teeth). Schenck (1928) has estimated the thickness of the Eugene Formation at Eugene to be near 5,000 feet. Hickman suggests the environmental setting was a shallow marine upper shelf away from open ocean water masses. She notes that during the Oligocene at this latitude a transition was taking place from tropical to more temperate faunas. Shallow embayments such as that of the Eugene Formation may have formed a protected environment that retained for a time the resident tropical forms while receiving the migrants from cooler waters, thus yielding a high diversity assemblage of mixed environmental types. Abundant molluscs in the Eugene Formation include species of *Nuculana*, *Acila*, *Tellina*, *Parvicardium*, *Diplodonta*, *Gemmula*, *Pitar*, *Bruclarkia*, *Crepidula*, *Solena*, and *Spisula*.

NEOGENE

In attempting to correlate Pacific Northwest units to California, Neogene units here have been variously assigned and reassigned to Oligocene, Miocene, Pliocene and Pleistocene. The solution to much of the confusion would be to use provincial stages, and Addicott (1976) has only recently proposed six molluscan stages to apply to the Pacific Northwest marine Neogene. These stages are all defined on the basis of molluscan ranges, and it is important to note that age assignments may be made by working with taxa at the generic level. As stated elsewhere, due to Tertiary sea withdrawal (regression) the last widespread Tertiary seas are Oligocene. Miocene and younger exposures are confined to a narrow coastal strip extending intermittently from Bandon/Cape

Blanco to Astoria, Oregon. River valleys cutting across this strip expose the section particularly well and one of the best of this type is the Yaquina River and estuary at Newport.

Oldest marine Miocene units in the Newport area are the upper Yaquina Formation and overlying Nye Mudstone. Addicott (Snavely, et al., 1964) correlates the Nye Mudstone with the Vaqueros of California in part on the basis of molluscs. Nye Mudstone molluscs are of a cool temperate variety from middle and outer neritic depths including the genera *Acila*, *Lucina*, *Nuculana*, *Yoldia*, *Thracia*, *Thyastria*, *Pecten*, *Tellina*, *Macoma*, *Pitar*, *Polinices*, *Cancellaria*, and *Siphonalia*. Addicott (1976) suggests the fossiliferous Yaquina Formation and possibly the lowermost Nye Mudstone may be correlated with Juanian (lowermost Miocene) stage.

By far the most widely distributed, best known marine Miocene unit in the State is the Astoria Formation. Although this type section is at Astoria, Oregon, best exposures are in a coastal strip in the vicinity of Newport and northward. Dodds (1963) has written an interesting article on the gradual covering of the type area at Astoria by the processes of street and building construction at the waterfront. Addicott (1976) has designated the "Newportian" Stage for molluscan rich exposures of the Astoria Formation where it crops out north of Newport to Fogarty Creek. The Astoria overlies unconformably the Yaquina and Nye Formations and attains a thickness of up to 2,000 feet in the Depoe Bay area. The unit consists of concretionary sandstones and siltstones with biostromal lenses of molluscs. First named the "Astoria Shales" by Condon, the Astoria Formation has received considerable attention even in recent years. Original reports of molluscs were made by Dana in the U. S. Exploring Expedition of 1838-1842 and described by Conrad in 1848 and Dall in 1909, but to date the most thorough treatise on the Astoria is by Ellen James Moore (1963). She has reported 97 species of molluscs from the unit as well as corals, brachiopods, bryozoa, and echinoids. Elsewhere, foraminifera and several vertebrate fossils have been described from the formation including pinnipeds, cetacea, fish and terrestrial vertebrates. Moore notes that all but 7 of the molluscan genera from the Astoria are still living off the Pacific Coast today. The fauna is dominantly a soft substrate non-rocky type. Common genera include *Anadara*, *Katherinella*, *Patiopecten*, and *Acila*. Most of the specimens show little or no transport, wear or abrasion. The abundance of articulated pelecypods or those with closed valves and the lack of

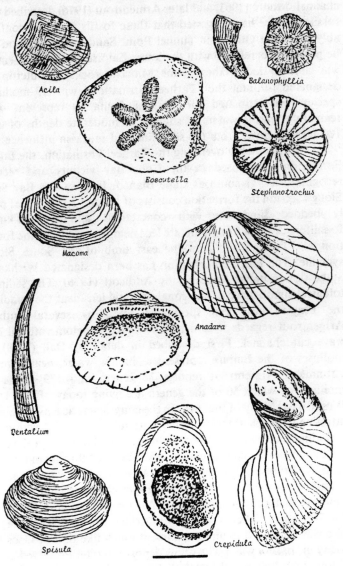

Acila

Balanophyllia

Eoscutella

Stephanotrochus

Macoma

Anadara

Dentalium

Spisula

Crepidula

wear suggest *in situ* burial. The populations are exclusively upper neritic, non-brackish types that lived in warm temperate water. Nomland (1917) has described five species of coral from the Astoria Formation.

Miocene rocks occur in the Coos Bay area but were unreported until 1949 when molluscan rich siltstones were recovered and piled up on a disposal area during a dredging operation of the Coos Bay channel. Moore (1963) and later Armentrout (1976) described these fossils. Moore has suggested that these fossils may rest conformably upon the Oligocene Tunnel Point Sandstone there and may be partially correlative with the lower Empire locally. Armentrout was able to locate these same Miocene rocks in outcrop and designated the unit the "Tarheel Formation" with 19 molluscan species. He estimated the Tarheel fauna to represent warm temperate marine waters of shallow to moderate depths of up to 180 feet but away from the shoreline and brackish influence.

Unconformably overlying the Tarheel Formation, the Empire Formation is exposed in the Coos Bay vicinity and recently recognized as far south as Cape Blanco. In the Coos Bay South Slough section the formation consists of around 3,000 feet of poorly bedded sandstone with concretions and a well-known fossiliferous Coos Conglomerate lens near the middle of the formation at "Fossil Point" on the east limb of the South Slough syncline. The Empire Formation has been designated Wishkahan Stage (lower upper Miocene) by Addicott (1976). The sedimentological significance of the conglomeratic lens near the middle of the Empire has been speculated upon by several authors. Armentrout regards it as a marine beach conglomerate fill of a wave cut channel. First described in detail by Dall (1909) the molluscs of the Empire represent a shallow water, nearshore environment. Armentrout reports 72 genera and 129 species of molluscs of which 50 of the genera are living today off the U. S. West Coast. Many of the shells in the conglomeratic lens show wear attesting to beach abrasion. Bone material from marine vertebrates including cetacea (whale) and pinnipeds (sea lions and seals) is not uncommon in the conglomerate lens. Scattered throughout the formation, bryozoa, brachiopods, crustacea and echinoderms are found in association with the prolific molluscan fauna. Common molluscs include: *Pseudocardium*, *Nuculana*, *Anadara*, *Crepidula*, *Chione*, *Natica*, *Pecten*, and *Solen*. Armentrout concludes the Empire was deposited in an environment much like that of Coos Bay today in shallow water with a low deposition rate. Preserved intact, colonies of the open "slipper shell", *Crepidula princeps*, indicate protected, quiet water such as in a coastal embayment. Molluscs found in the Empire reflect a temperate to warm temperate climate during this interval.

The effect of Berggren and VanCouvering's (1974) reassigning the base of the Pliocene from 10 million up to 5 million years b.p.

Macrocallista

Solen

Psephidia

Nucula

Chione

Mya

Mytilus

Panope

Aturia

1"

Saxidomus

has been to remove the Empire Formation from the Pliocene and place it in the upper Miocene. In his designation of Neogene molluscan stages in the Pacific Northwest, Addicott (1976)

111

establishes the Pliocene Moclipsian Stage in fossiliferous exposures of the Quinault Formation in Washington State but does not note any fossiliferous Pliocene marine in Oregon.

Baldwin (1945) has designated the term "Port Orford Formation" for sediments which lie between the Empire Formation and the Elk River Beds at Cape Blanco. Bounded above and below by unconformities, the Port Orford Formation grades from coarse sand and conglomerate near its base to argillaceous (shale-rich) sand near the top bearing fossiliferous concretions. Although detailed paleontology of the Port Orford has not been attempted, stratigraphic position would suggest that they are latest Pliocene in age.

The Elk River Beds of Diller (1902) exposed at Cape Blanco have been variously regarded by authors as Pliocene and Pleistocene. Addicott (1964) in a paper on Cape Blanco Pleistocene states that the name "Elk River Beds" should be restricted to gently dipping fossiliferous siltstones and sandstones near the base of the seacliffs and not to upper Pleistocene terrace deposits. These units are of probable late Pliocene age and bear a diverse molluscan fauna dominated by the clam, *Psephidia*, and indicator of paleodepths of less than 30 meters. Four distinct units crop out at Cape Blanco and south along the sea cliff exposures to the mouth of the Elk River. At the Cape, dipping Empire Formation is overlain by Pleistocene terrace deposits. Toward the South, the Empire is overlain by the Port Orford Formation which is in turn overlain by the Elk River Beds. The latter are nearly horizontal and often are difficult to distinguish from the overlying Pleistocene terrace unit. Because of the excellent cross-sectional exposures and rich clam fauna, Clifton and Boggs (1970) chose the Elk River Beds as a study site for the effects of waves upon clam shells after death but prior to final burial. These authors were able to show that the classic "concave-side-down" configuration for clam shells may well be disturbed by migrating ripples overturning the shells. This geopedal feature for distinguishing the top and bottom of folded geologic units may then only be correct half the time.

Pleistocene shell-bearing fossiliferous terrace deposits are known from only a few localities along the Oregon coast. Probably the best known in this category are the seacliff exposures at Cape Blanco. Several authors have published on these, but the best summary to date on the molluscs is that of Addicott (1964). He reports an upper Pleistocene current/wave disturbed, sandy, inter-sublittoral molluscan dominated fauna of 12 clam species, a par-

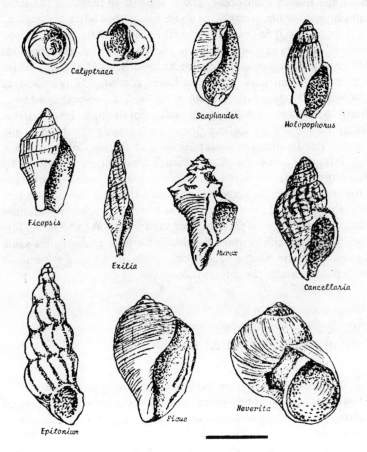

Calyptraea

Scaphander

Molopophorus

Ficopsis

Exilia

Murex

Cancellaria

Epitonium

Ficus

Neverita

ticularly diverse assemblage of 24 gastropods, 5 barnacles and assorted echinoid and bryozoan debris as well as foraminifera. Although wave action may have disturbed the molluscan shells in what may have been a very slow rate of deposition, abundant articulated valves of the large clams *Saxidomus* and *Tresus* suggest that the fossil material was not transported any considerable distance. The fauna is decidedly a cool water type as 20 of the fossil species are today known only from areas considerably north of the present site of Cape Blanco.

Another Pleistocene mollusc locality is noted by Baldwin (1950) from near Newport in the Coquille Formation exposed in a small cove north of the Newport jetty. Molluscs at this locality include *Thais*, *Tresus* and *Macoma*. Most of these forms co-occur in the Cape Blanco Pleistocene, and it appears to represent the same environment if not roughly the same age as that latter occurrence.

Zullo (1969) has examined fossiliferous pockets of Pleistocene terrace deposits at Coquille Point and Grave Point near Bandon. He reports high concentrations of material with 58 varieties of invertebrates dominated by the molluscs *Hiatella*, *Macoma*, *Mya*, *Psephidia*, *Saxidomus*, *Fusitriton*, as well as several crustacea, bryozoa, echinoderms, brachiopods, corals and foraminifera. Many of the species Zullo reports are presently restricted to the coastal Pacific north of Puget Sound and British Columbia. This suggests that, like the Cape Blanco assemblage described by Addicott (1964), these fossils represent a substantially cooler environment than exists today at the same latitude. Zullo summarizes that the Bandon fauna was from the lower intertidal and subtidal limits of a rocky coast that was protected from wave shock. He further notes that although the Bandon material is probably the same age as the Cape Blanco fauna, only 30% of the species occur in both places due to the different habitats.

FRESHWATER AND TERRESTRIAL (AIR BREATHING) MOLLUSCS

Fossils of non-marine molluscs occur in several localities in Oregon but never in anywhere near the diversity and numbers of marine types. Non-marine snails and clams are far less common in the environment than their marine counterparts, but their limited preservability also contributes to their poor fossil record. Terrestrial or air breathing snails bearing a dorsal shell lack the buoyancy advantage of the aquatic environment and consequently develop a thin shell that deteriorates easily. Such shells, for example, can be crushed readily between the fingers like an eggshell, whereas any given marine snail shell would require considerable strength to crush similarly. Freshwater clams, by contrast, are often thicker-shelled than marine forms, but the former may be composed almost wholly of the mineral aragonite, whereas the latter are largely calcite. With time and burial, the aragonite shell, or "mother of pearl", is unstable and tends to break down to reform the more stable calcium carbonate mineral calcite. In this recrystallization process the volume change and resultant fractures in the shell make it particularly subject to solution by percolating ground water. Often, even if the shell remains intact in the rock, the effect of fracturing accompanying recrystallization causes it to crumble as it is removed from the entombing rock matrix.

Only a few terrestrial and freshwater molluscan fossils are described in the State, and many of these are from the John Day beds where they occur in association with mammalian fossils. Some of the earliest work in this regard was by White (1885) and Stearns (1902, 1906), wherein they describe terrestrial or air breathing land snails from the John Day and Mascall Formations. Included in this work was the assessment of modern forms in the region today. Nonmarine molluscs evolve slowly, and most of the genera described by the above authors exist today in the same region including *Unio, Pyramidula, Polygyra, Ammonitella, Helix, Lymnea*, and *Epiphragmorphora*. In a summary paper on the John Day nonmarine molluscs, G. Dallas Hanna (1920) noted the very high frequency of certain genera, for example, *Polygyra*, and the affinity to asiatic types, *Rhiostoma*. In a similar review paper Hanna (1922) described nonmarine molluscs from possible eastern Oregon Eocene sediments.

Middle Miocene (Clarendonian) freshwater pelecypods and gastropods are listed by Taylor (1963) from western Malheur and Harney Counties where they occur in association with Shotwell's Juntura Basin Black Butte local mammalian fauna. Some three species of pelecypods and six species of gastropods suggest a perennial, freshwater lake environment with 15-30 feet of water. Commonest genera occuring in three separate localities include *Sphaerium, Fluminicola, Viviparus, Radix*, and *Carinifex*.

In a study of the Yonna Formation of the Klamath River Basin (Newcomb, 1958), freshwater molluscs were examined by Hanna and Ten-Chien Yen. This limited fauna included species of the genera *Sphaerium, Valvata, Amnicola, Lanx, Physa, Vorticifex*, and *Lymnea*, and both authors concluded that the parent rock of the Yonna Formation was of Pliocene age.

Taylor (1960) in a short paper on the distribution of the living northeastern California, southcentral Oregon freshwater clam, *Pisidium ultramontauum*, regards that species as a relict form of a much wider Pliocene/Pleistocene distribution. Although the present distribution of living forms is somewhat limited, the occurrence of fossil forms of this species extends over a much wider area as far as eastern Idaho. Taylor regards this species of *Pisidium* along with species of the snail, *Carinifex*, and the freshwater fish, *Chasmistes* (sucker), as part of the biologic signature of a previous-

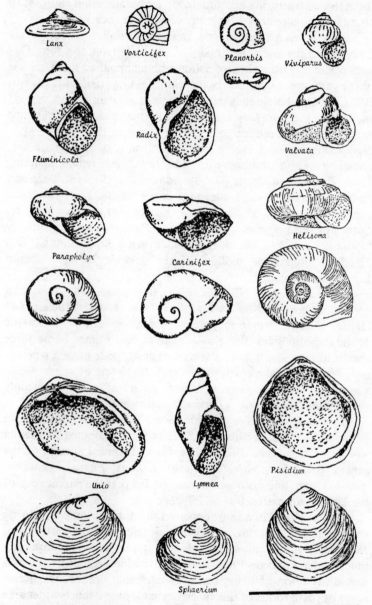

Lanx

Vorticifex

Planorbis

Viviparus

Fluminicola

Radix

Valvata

Parapholyx

Carinifex

Helisoma

Unio

Lymnea

Pisidium

Sphaerium

ly connected chain of lakes and drainage basins that extended for hundreds of miles in two main branches into northern California and Nevada as well as across Oregon and southern Idaho. He concluded from these findings that the present day course of the Snake River is relatively young and that at one time the Snake River flowed into the Pacific Ocean independent of the Columbia and Sacramento drainage systems.

Although Pliocene localities in the State are very limited, freshwater snails from late Tertiary Summer Lake beds in west central Lake County are regarded as Pliocene by Hanna (1922; 1963). Vertebrate mammalian faunas from the same area are considered to be Pleistocene (Hay, 1927). Genera from the Summer Lake fauna include *Parapholyx*, *Planorbis*, *Lymnea*, *Vorticifex*, *Helisoma*, *Gyraulis*, *Lymnophysa*, and *Valvata*. Freshwater molluscs are often listed in the course of reporting the geology or in describing fossil groups of birds or fish.

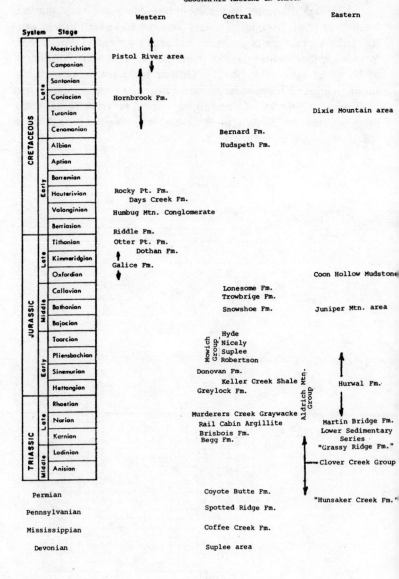

OREGON PRECENOZOIC INVERTEBRATE FOSSIL BEARING FORMATIONS

GEOGRAPHIC REGIONS IN OREGON

System		Stage		Western	Central	Eastern
CRETACEOUS	Late	Maestrichtian		Pistol River area		
		Campanian				
		Santonian				
		Coniacian		Hornbrook Fm.		Dixie Mountain area
		Turonian				
		Cenomanian			Bernard Fm.	
		Albian			Hudspeth Fm.	
	Early	Aptian				
		Barremian				
		Hauterivian		Rocky Pt. Fm. / Days Creek Fm.		
		Valanginian		Humbug Mtn. Conglomerate		
		Berriasian		Riddle Fm.		
JURASSIC	Late	Tithonian		Otter Pt. Fm. / Dothan Fm.		
		Kimmeridgian		Galice Fm.		
		Oxfordian				Coon Hollow Mudstone
	Middle	Callovian			Lonesome Fm. / Trowbrige Fm.	
		Bathonian			Snowshoe Fm.	Juniper Mtn. area
		Bajocian				
	Early	Toarcian			Hyde / Nicely / Suplee / Robertson	
		Pliensbachian				
		Sinemurian			Donovan Fm. / Keller Creek Shale	Hurwal Fm.
		Hettangian			Greylock Fm.	
TRIASSIC	Late	Rhaetian			Murderers Creek Graywacke	
		Norian			Rail Cabin Argillite	Martin Bridge Fm.
		Karnian			Brisbois Fm. / Begg Fm.	Lower Sedimentary Series / "Grassy Ridge Fm."
	Middle	Ladinian				Clover Creek Group
		Anisian				

Mowich Group: Hyde, Nicely, Suplee, Robertson

Aldrich Mtn. Group

Permian   Coyote Butte Fm.   "Hunsaker Creek Fm."

Pennsylvanian   Spotted Ridge Fm.

Mississippian   Coffee Creek Fm.

Devonian   Suplee area

118

The following are fossil invertebrate ranges
listed in the published literature.

	Curry Co. Cretaceous	Hornbrook Fm.	Dixie Mtn.	Bernard Fm.	Hudspeth Fm.	Days Creek Fm.	Rocky Point Fm.	Humbug Mtn. Congl.	Riddle Fm.	Otter Point Fm.	Dothan Fm.	Galice Fm.	Coon Hollow Mudst.	Lonesm./Trowbr. Fm.	Snowshoe Fm.	Juniper Mtn.	Hyde/Nicely Fm.	Suplee/Robertson Fm.	Donovan Fm.	Graylock Fm.	Pail Cabin Argill.	Hurwal/Martin Br.	Brisbois/Begg Fm.	"Grassy Ridge Fm."	Clover Creek Grnst.	Husaker Cr Coyote Butte	Coffee Creek Fm.	Suplee area Devonian
Acanthodiscus a					●																							
Acompsoceras a	●																											
Acriocerns a					●																							
Acteonina g																										●		
Alexenia b																										●		
Ambocoelia b				●																								
Ampullina g	●																											
Anagaudryceras a	●																											
Anatina p																			●									
Anidanthus b																										●		
Anisoceras a	●																											
Anomia p				●																								
Anthonya p				●																								
Antiquatonia b																										●		
Aphrodina p																				●	●	●						
Arcestes a																	●											
Arcomya p																		●										
Arieticeras a																●	●	●										
Astarte p																●	●											
Asthenoceras a																●	●											
Astrocoenia c																	●											
Atractites																												●
Atrypa b									●																			
Aulacoteuthis be																			●									
Arnioceras a																										●		
Avonia p																										●		
Botostomella br																												●
Batterobyia c																												●
Belodella co					●																							
Beudanticeras a		●																										
Bostrychoceras a																			●									
Bisiphytes n		●																										
Brancoceras a		●																										
Brewericeras a				●																								
Buchia p							●		●	●	●	●																
Calliphylloceras a	●																		●									
Calycoceras a		●																										
Camarophoria b																									●	●		
Campophyllum b																											●	
Camptonectes p																	●											
Canavaria a																			●	●								
Cardinia p																				●								
Cardioceras a													●															
Cardium p	●																											
Cassianella p																								●				
Catacoeloceras a																●	●											
Catulloceras a																●												
Cercomya p																					●							
Chalamys p																●												
Chondroceras a																●												
Chonetinella b																										●		
Chonetes b																										●		

119

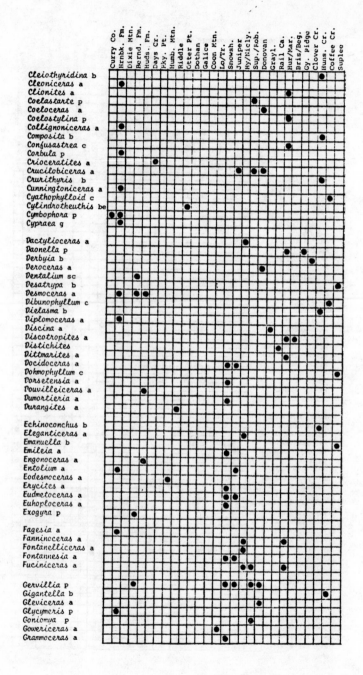

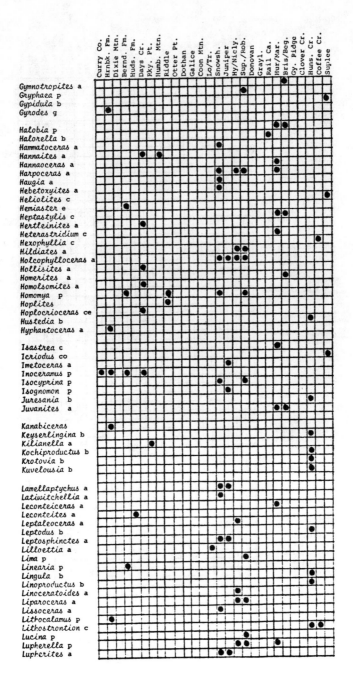

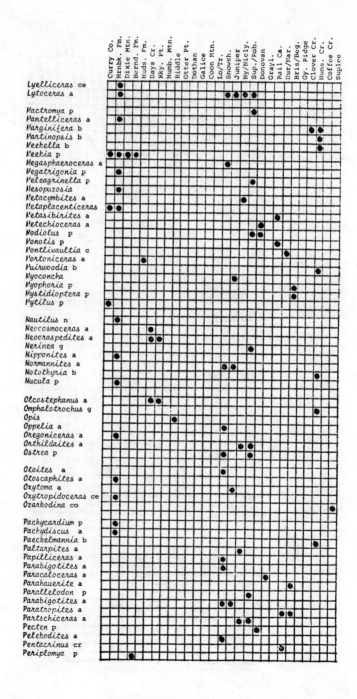

	Curry Co.	Hrnbk. Fm.	Dixie Mtn.	Bernd. Fm.	Huds. Fm.	Days Cr.	Rky. Pt.	Humb. Mtn.	Riddle	Otter Pt.	Dothan	Galice	Coon Mtn.	Lo/Tr.	Snowsh.	Juniper	Hy/Nicly.	Sup./Rob.	Donovan	Grayl.	Rail Ca.	Hur/Xar.	Bris/Beg.	Gy. Ridge	Clover Cr.	Huns. Cr.	Coffee Cr.	Suplee
Pervinquieria a	●																											
Phacoides p															●													
Pholadomya p																	●	●										
Phylloceras a								●	●	●					●	●	●	●										
Phymatoceras a															●													
Pinna p	●			●													●	●		●								
Placerites																												
Planammatoceras a															●													
Pleurohorridonia b																								●				
Pleuromya p	●														●	●	●	●										
Pleydellia a															●	●												
Plicatostylus p	●																											
Plicatula p																												
Poecilomorphus a																												●
Polygnathus co																												
Polyplectus a																												
Polyptychites a						●		●																				
Posidonia p															●	●												
Praestrigites a															●	●												
Prionocyclus a	●																											
Prionotropis a	●																											
Probolonia b																								●				
Protoscidella b																								●				
Procycldonautilus ce																	●					●						
Prodactylioceras a																	●	●										
Productus br																								●	●			
Proniceras a							●		●																			
Protacanthodiscus a								●	●																			
Protogrammoceras a																	●	●				●						
Pseudomartinia b																				●								
Psiloceras a	●																											
Pteria p	●	●																										
Pterotrigonia p	●	●																										
Pugnellus g																												
Punctospirfer b																								●				
Purpuroidea p																												
Pustula b																												
Pusozigella a				●																								
Remondia p	●																											
Reynesoceras a																	●	●						●				
Rhipidomella b																								●				
Rhynchonella b																●	●									●		
Rhynchopora b																										●		
Rostranteris b																										●		
Sagenites a																												
Sandlingites a																												
Sarasinella a				●	●	●																						
Scalarites a	●																											
Scaphites a	●																											
Schizophoria b																												●
Shasticrioceras a	●					●																						
Skirroceras a																												
Simbirskites a																												
Sirenites ce																				●								
Solecurtus	●																											
Sonninia a															●													
Speetoniceras a						●																						
Sphaeoceras a															●													
Spinatrypa b																										●	●	
Spirifer b																									●			
Spiriferella b																										●		
Spiriferina b																●	●											
Spiroceras a							●																					
Spiticeras a							●																					

123

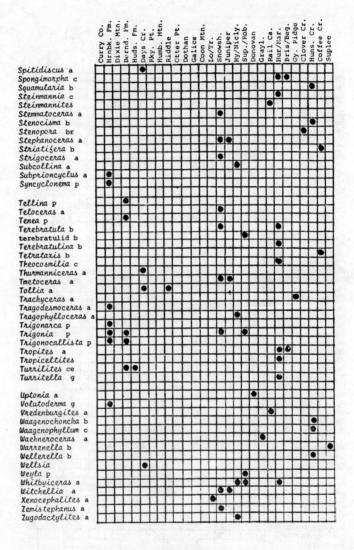

a-ammonite .
b-brachiopod
be-belemnite
br-bryozoa
c-coral
ce-cephalopod
co-conodont

cr-crinoid
e-echinoid
g-gastropod
n-nautiloid
p-pelecypod
sc-scaphopod

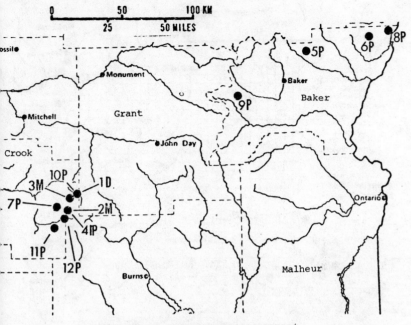

OREGON FOSSIL INVERTEBRATE LOCALITIES

Following is a list of Oregon fossil invertebrate localities. The reference numbers are to authors listed in the
bibliography who have published on the specific localities.

DEVONIAN
   1. Suplee area (Berger Ranch), Crook Co. Refs. 94, 101, 178,

MISSISSIPPIAN
   2. Coffee Creek valley, @ 10 mi. SW of Suplee, Crook Co.
      Coffee Creek Fm. Refs. 119, 130a, 131.
   3. Trout Creek valley, @ 3 mi. SW of Suplee, Crook Co. Coffee
      Creek Fm. Refs. 130a, 176.

PENNSYLVANIAN
   4. Spotted Ridge, @ 10 mi. SW of Suplee (Mills Ranch), Crook Co.
      Spotted Ridge Fm. Refs. 119, 131.

PERMIAN
   5. Clover Creek, Baker Co. Clover Creek Group. Permian/Triassic.
      Refs. 67, 193.
   6. Cornucopia (@ 5 mi. E of), Baker Co. No Fm. Refs. 231.
   7. Grindstone Creek, Crook Co. Coyote Butte Fm. Refs. 40, 130a.
   8. Homestead, Baker Co. "Hunsaker Creek Fm." Refs. 193, 219.
   9. Sumpter, Baker Co. ?Elkhorn Ridge Argillite. Refs. 246.
  10. Suplee area, Crook Co. Coyote Butte Fm. Refs. 40.
  11. Tuckers Butte, Crook Co. Coyote Butte Fm. Refs. 40, 130a, 131.
  12. Twelvemile Creek, Crook Co. Coyote Butte Fm.
      Refs. 40, 130a.

125

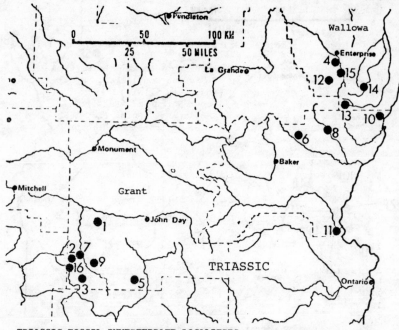

TRIASSIC FOSSIL INVERTEBRATE LOCALITIES

1. Aldrich Mtn., Grant Co. Aldrich Mountain Group. Triassic/Jurassic. Refs. 26.
2. Begg Creek valley, Grant Co. Begg Fm. Triassic. Refs. 48.
3. Big Flat, @ 5 mi. SE of Izee, Grant Co. Brisbois Fm. Triassic. Refs. 48, 180a.
4. Black Marble Quarry, Wallowa Co. Martin Bridge Fm. Triassic. Refs. 193.
5. Camp Creek valley, Grant Co. Brisbois Fm. Triassic. Refs. 48.
6. Clover Creek, Baker Co. Clover Creek Group. Permian/Triassic. Refs. 67, 193.
7. Cow Creek valley, Grant Co. Brisbois Fm. Triassic. Refs. 48.
8. Eagle Creek valley, Baker Co. No Fm. Triassic. Refs. 192, 193.
9. Elkhorn Creek Canyon, Grant Co. Rail Cabin Argillite. Triassic. Refs. 48.
10. Homestead, Baker Co. ?Grassy Ridge Fm. Triassic. Refs. 219.
11. Huntington, Baker Co. No Fm. Triassic. Refs. 12f, 16.
12. Hurricane Divide, Wallowa Co. Martin Bridge Fm./Hurwal Fm. Triassic/Jurassic. Refs. 193.
13. Imnaha River valley (south fork), Wallowa Co. Hurwal Fm. Triassic/Jurassic. Refs. 193.
14. Joseph (@ 10 mi. SE of), Wallowa Co. Hurwal Fm. Triassic/Jurassic. Refs. 193.
15. Point Joseph, Wallowa Co. Lower Sedimentary Series/Martin Bridge Fm. Triassic. Refs. 193.
16. Suplee area, Grant Co. Brisbois Fm. Triassic. Refs. 48.

126

JURASSIC FOSSIL INVERTEBRATE LOCALITIES

1. Aldrich Mtn., Grant Co. Aldrich Mountain Group. Triassic/
   Jurassic. Refs. 26.
2. Beaver Creek valley, Grant Co. Mowich Group/Snowshoe Fm.
   Refs. 48, 89, 89a, 89b, 117.
3. Brogan (@ 6 mi. N of), Malheur Co. No Fm. Refs. 89b, 225.
4. Buck Creek, @ 10 mi. NW of Riddle, Douglas Co. Myrtle Group.
   Jurassic/Cretaceous. Refs. 49e, 91, 104.
5. Cape Blanco, Curry Co. ?Otter Point Fm. Refs. 104.
6. Chetco River, Curry Co. ?Dothan Fm. Refs. 12d, 166.
7. Clover Creek valley, 10 mi. SE of Ironside, Malheur Co.
   No Fm. Refs. 89b.
8. Collier Butte, Curry Co. ?Otter Point Fm. Refs. 104.
9. Cow Creek, Grant Co. Mowich Group. Refs. 48.
10. Crook Point (5 mi. S of), Curry Co. Otter Point Fm.
    Refs. 104.
11. Days Creek, Douglas Co. Myrtle Group. Jurassic/Cretaceous.
    Refs. 90.
12. Dillard area, Douglas Co. Myrtle Group. Jurassic/Cretaceous.
    Refs. 44e, 91.
13. Elk River, Curry Co. Myrtle Group. Jurassic/Cretaceous.
    Refs. 90, 104, 164.
14. Enterprise (5 mi. SW of at Sheep Ridge), Wallowa Co.
    Hurwal Fm. Refs. 89, 89a.
15. Euchre Creek valley, Curry Co. Otter Point Fm. Refs. 104, 90.
16. Flat Creek, Grant Co. Snowshoe/Trowbridge Fm. Refs. 48,
    89b, 117.
17. Freeman Creek, Grant and Harney Co. Mowich Group/Snowshoe Fm.
    Refs. 48, 89b.
18. Grindstone Creek, Harney Co. Snowshoe Fm. Refs. 89b.
19. Humbug Mtn., Curry Co. Otter Point Fm. Refs. 104.
20. Hunter Creek, Curry Co. Otter Point Fm. Refs. 104.
21. Huntington (13 mi. W of), Baker Co. No Fm.
    Refs. 89b, 225.
22. Hurricane Divide, Wallowa Co. Martin Bridge/Hurwal Fm.
    Triassic/Jurassic. Refs. 193.
23. Imnaha River (N of mouth, Snake River Canyon), Wallowa Co.
    Coon Hollow Mudstone. Refs. 144.
24. Imnaha River valley (south fork), Wallowa Co. Hurwal Fm.
    Triassic/Jurassic. Refs. 193.
25. Ironside Mtn. (10 mi. SW of), Malheur Co. No Fm.
    Refs. 89b, 225.
26. Izee area, Grant Co. Mowich Group/Snowshoe Fm./Lonesome Fm.
    Refs. 48, 89, 89a, 89b, 117.
27. Johnson Creek valley, Coos Co. Galice Fm. Refs. 49e, 51,
    104.
28. Joseph (10 mi. SE of), Wallowa Co. Hurwal Fm. Triassic/
    Jurassic. Refs. 193.
29. Juniper Mtn., Malheur Co. No Fm. Refs. 62, 89b, 225.
30. Morgan Creek valley, Grant Co. Graylock Fm. Refs. 48.
31. Myrtle Creek valley, Douglas Co. Myrtle Group. Jurassic/
    Cretaceous. Refs. 49e, 91, 104, 164.
32. Otter Point, Curry Co. Otter Point Fm. Refs. 104.
33. Port Orford area, Curry Co. Otter Point Fm. Refs. 104.
34. Riddle (12 mi. W of), Douglas Co. Myrtle Group. Jurassic/
    Cretaceous. Refs. 90, 91.
35. Rosebud Creek, Grant Co. Mowich Group/Snowshoe Fm.
    Refs. 48, 117.

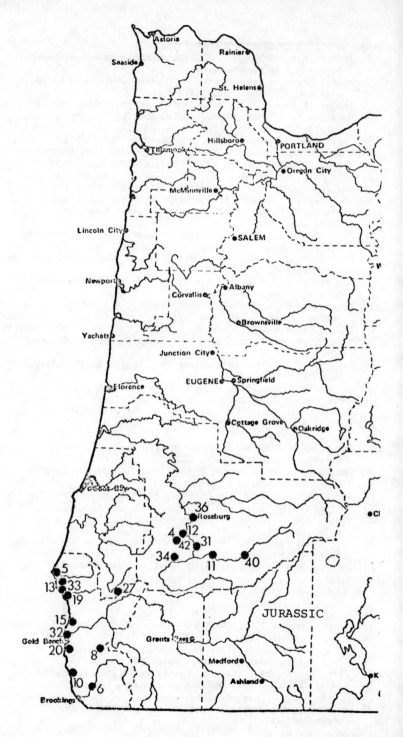

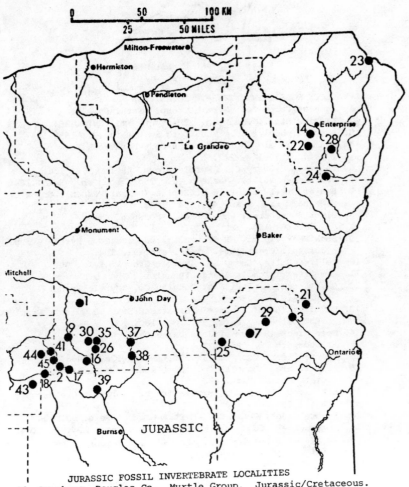

CRETACEOUS FOSSIL INVERTEBRATE LOCALITIES
1. Agness, Curry Co. Myrtle Group. Refs. 90, 164.
2. Aldrich Mtns., Grant Co. ?Hudspeth Fm. Refs. 164.
3. Antone, Wheeler Co. ?Hudspeth Fm. Refs. 95, 155, 164.
4. Ashland (3 mi. S of), Jackson Co. Hornbrook Fm.
      Refs. 157, 164, 227.
5. Bald Mtn., Curry Co. Myrtle Group. Refs. 104, 164.
6. Bear Creek valley, Jackson Co. ?Hornbrook Fm. Refs. 164.
7. Beaver Creek valley, N of Suplee (Bernard Ranch), Crook Co.
      Bernard Fm. Refs. 48, 164, 180b.
8. Black Butte area, Wheeler Co. Hudspeth Fm.
      Refs. 48, 164, 180b, 239.
9. Buck Peak, 10 mi. NW of Riddle, Douglas Co. Myrtle Group.
      Jurassic/Cretaceous. Refs. 49e, 91, 104.
10. Collier Butte, Curry Co. Myrtle Group. Refs. 90.
11. Cow Creek (Nichols Station)Douglas Co,Myrtle Group. 90, 91.
12. Days Creek, E of Riddle, Douglas Co. Myrtle Group.
      Jurassic/Cretaceous. Refs. 91, 164.
13. Dayville (SW of at Battle Creek and Spanish Gulch), Grant Co.
      ?Hudspeth Fm. Refs. 95, 164, 180.
14. Deer Creek (S fork of John Day River), Grant Co.
      ?Hudspeth Fm. Refs. 164.
15. Dement Creek, Coos Co. No Fm. Refs. 12e, 164.
16. Dillard area, Douglas Co. Myrtle Group. Jurassic/
      Cretaceous. Refs. 49e, 91.
17. Elk River, Curry Co. Myrtle Group. Jurassic/Cretaceous.
      Refs. 90, 104, 164.
18. Euchre Creek, Curry Co. Myrtle Group. Refs. 90, 104.
19. Foggy Creek, Coos Co. Myrtle Group. Refs. 12d, 91, 164.
20. 49 Placer Mine, 5 mi. SW of Medford. Hornbrook Fm.
      Refs. 5, 95, 118, 164.

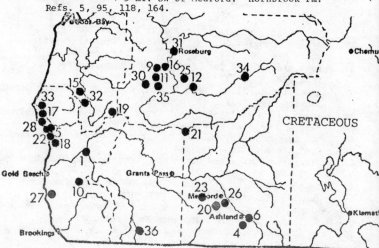

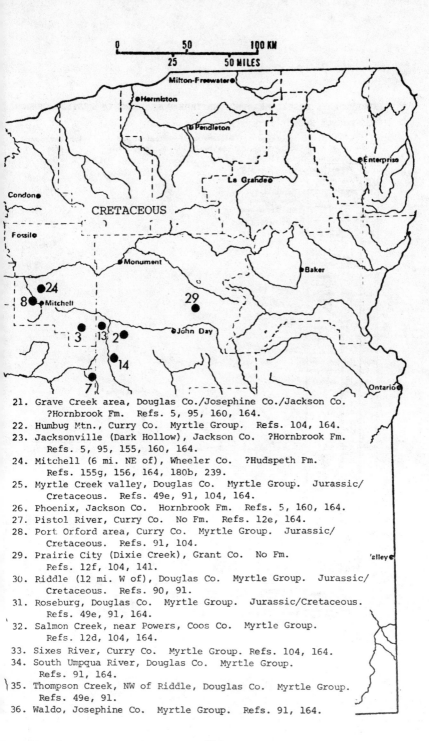

CRETACEOUS

21. Grave Creek area, Douglas Co./Josephine Co./Jackson Co.
    ?Hornbrook Fm. Refs. 5, 95, 160, 164.
22. Humbug Mtn., Curry Co. Myrtle Group. Refs. 104, 164.
23. Jacksonville (Dark Hollow), Jackson Co. ?Hornbrook Fm.
    Refs. 5, 95, 155, 160, 164.
24. Mitchell (6 mi. NE of), Wheeler Co. ?Hudspeth Fm.
    Refs. 155g, 156, 164, 180b, 239.
25. Myrtle Creek valley, Douglas Co. Myrtle Group. Jurassic/
    Cretaceous. Refs. 49e, 91, 104, 164.
26. Phoenix, Jackson Co. Hornbrook Fm. Refs. 5, 160, 164.
27. Pistol River, Curry Co. No Fm. Refs. 12e, 164.
28. Port Orford area, Curry Co. Myrtle Group. Jurassic/
    Cretaceous. Refs. 91, 104.
29. Prairie City (Dixie Creek), Grant Co. No Fm.
    Refs. 12f, 104, 141.
30. Riddle (12 mi. W of), Douglas Co. Myrtle Group. Jurassic/
    Cretaceous. Refs. 90, 91.
31. Roseburg, Douglas Co. Myrtle Group. Jurassic/Cretaceous.
    Refs. 49e, 91, 164.
32. Salmon Creek, near Powers, Coos Co. Myrtle Group.
    Refs. 12d, 104, 164.
33. Sixes River, Curry Co. Myrtle Group. Refs. 104, 164.
34. South Umpqua River, Douglas Co. Myrtle Group.
    Refs. 91, 164.
35. Thompson Creek, NW of Riddle, Douglas Co. Myrtle Group.
    Refs. 49e, 91.
36. Waldo, Josephine Co. Myrtle Group. Refs. 91, 164.

INVERTEBRATE FOSSIL BEARING TERTIARY FORMATIONS OF WESTERN OREGON

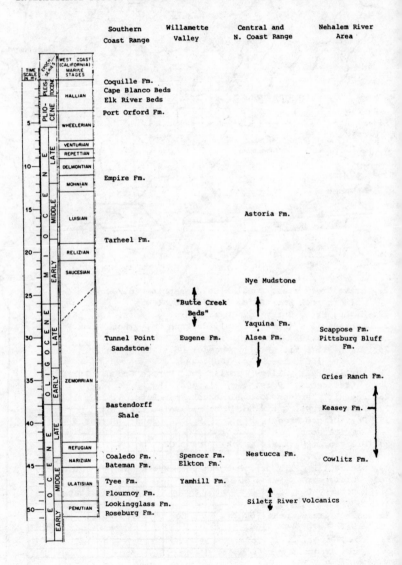

	Southern Coast Range	Willamette Valley	Central and N. Coast Range	Nehalem River Area
HALLIAN	Coquille Fm. Cape Blanco Beds Elk River Beds Port Orford Fm.			
WHEELERIAN				
VENTURIAN				
REPETTIAN				
DELMONTIAN				
MOHNIAN	Empire Fm.			
LUISIAN			Astoria Fm.	
RELIZIAN	Tarheel Fm.			
SAUCESIAN			Nye Mudstone	
		"Butte Creek Beds"	Yaquina Fm.	
ZEMORRIAN	Tunnel Point Sandstone	Eugene Fm.	Alsea Fm.	Scappose Fm. Pittsburg Bluff Fm.
				Gries Ranch Fm.
	Bastendorff Shale			Keasey Fm.
REFUGIAN				
NARIZIAN	Coaledo Fm. Bateman Fm.	Spencer Fm. Elkton Fm.	Nestucca Fm.	Cowlitz Fm.
ULATISIAN	Tyee Fm. Flournoy Fm.	Yamhill Fm.		
PENUTIAN	Lookingglass Fm. Roseburg Fm.		Siletz River Volcanics	

132

OREGON CENOZOIC INVERTEBRATE RANGES

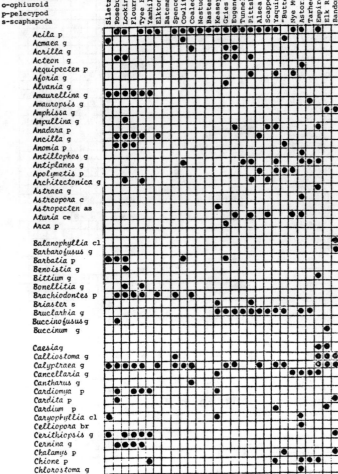

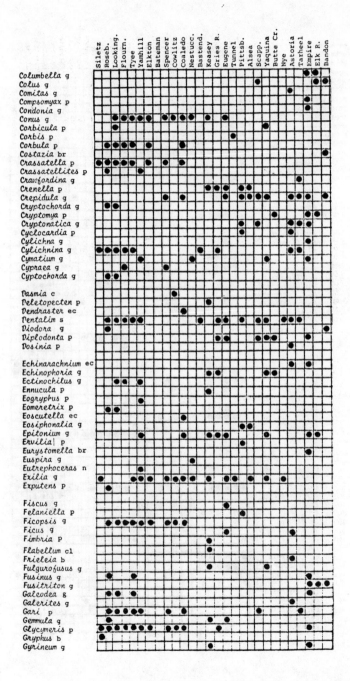

Columbella g
Colus g
Comitas g
Compsomyax p
Condonia g
Conus g
Corbicula p
Corbis p
Corbula p
Costazia br
Crassatella p
Crassatellites p
Crawfordina g
Crenella p
Crepidula g
Cryptochorda g
Cryptomya p
Cryptonatica g
Cyclocardia p
Cylichna g
Cylichnina g
Cymatium g
Cypraea g
Cyptochorda g

Dasmia c
Deletopecten p
Dendraster ec
Dentalim s
Diodora g
Diplodonta p
Dosinia p

Echinarachnium ec
Echinophoria g
Ectinochilus g
Ennucula p
Eogryphus p
Eomeretrix p
Eoscutella ec
Eosiphonalia g
Epitonium g
Ervilia| p
Eurystomella br
Euspira g
Eutrephoceras n
Exilia g
Exputens p

Fiscus g
Felaniella p
Ficopsis g
Ficus g
Fimbria p
Flabellum cl
Frieleia b
Fulgurofusus g
Fusinus g
Fusitriton g
Galeodea g
Galerites g
Gari p
Gemmula g
Glycymeris p
Gryphus b
Gyrineum g

Siletz, Roseb., Looking., Flourn., Tyee, Yamhill, Elkton, Bateman, Spencer, Cowlitz, Coaledo, Nestucc., Bastend., Keasey, Gries R., Eugene, Tunnel, Pittsb., Alsea, Scapp., Yaquina, Butte Cr., Nye, Astoria, Tatheel, Empire, Elk R., Bandon

	Siletz	Roseb.	Looking.	Flourn.	Tyee	Yamhill	Elkton	Bateman	Spencer	Cowlitz	Coaledo	Nestucc.	Bastend.	Keasey	Gries R.	Eugene	Tunnel	Pittsb.	Alsea	Scapp.	Yaquina	Butte Cr.	Nye	Astoria	Tarheel	Empire	Elk R.	Bandon
Halonanus p			●																									
Hemipleurotoma g										●	●																	
Hemithyris b																												●
Heteropora br	●																											●
Hiatella p																												
Hinia g																							●	●				
Homalopoma g		●	●																									
Isocrinus ec														●					●									
Katharina ch																												●
Katharinella p														●														
Kewia ec																●												
Knefastia g														●		●												
Lacuna g																										●		
Latirus g		●	●																								●	●
Lepeta g																												●
Leptophyllastrea cl	●																											
Leptoseris cl	●																											
Lima p														●														
Limopsis p																												
Liomesus g																												
Liracassis g																					●		●					
Lithia																	●											
Litorhadia p															●	●												
Littorina g																							●					●
Lophelia c																												
Lora g																												
Loxocardium p		●	●													●												
Loxotrema g																												
Lucina p	●	●	●									●					●	●			●			●	●			
Lucinoma p																												
Lyria g		●	●																									
Macoma p						●										●	●	●					●	●	●	●	●	●
Macrocallista p	●	●	●									●		●	●	●	●	●	●									
Malletia p																												
Mangelia p																									●	●		
Margarites g																									●	●		●
Martesia p																												
Mathildia g	●																											
Mediagro g																								●				
Megasurcula g																								●				
Megistostoma g			●	●																								●
Melanella g																												
Membranipora br																												
Mercimonia p		●	●	●																								
Microcallista		●		●																								
Minormalletia p													●															
Miopleiona g																												
Mitra g		●																										
Mitrella g														●														
Modiolus p		●				●				●	●					●							●	●	●	●		
Molophphorus g										●	●	●	●															
Nopalia ch																												●
Murex g		●	●																					●	●	●	●	
Mya p																							●	●		●		
Myadesma ch																●												
Mytilus p						●									●	●	●	●	●				●		●	●	●	●
Nassarius g																			●			●		●				
Natica g														●	●						●		●					
Nautilus n				●																								
Nehalemia g													●															
Nekewis g												●				●												
Nemocardium p	●	●	●	●	●		●			●			●		●	●												

135

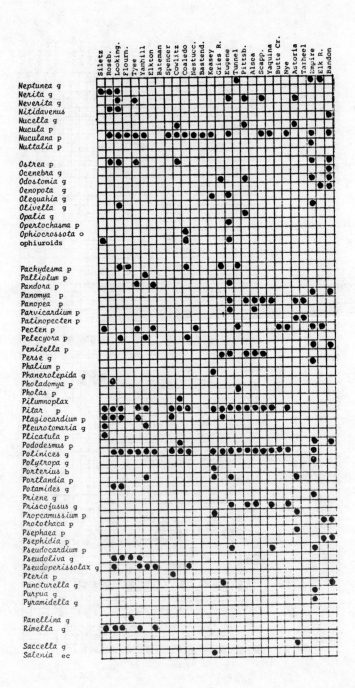

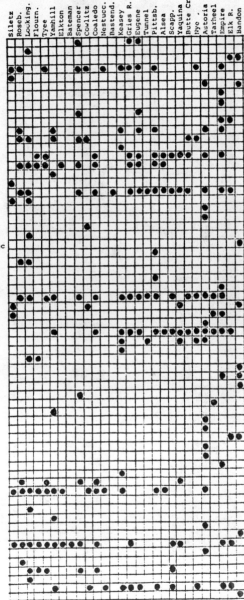

Sanguinolaria g
Sassia g
Saxiodomus p
Scaphander g
Schedocardia p
Searlesia g
Securella p
Schutellaster ec
Semele p
Siliqua p
Sinum g
Siphonalia g
Solemya p
Solen p
Solena p
Spatangus ec
Spiroglyphus g
Spisula g
Spondylus p
Stephanocyathus c
Stephanotrochus cl
Stichopsammia c
Streptochetus g
Strongylocentrotus ec
Suavodrillia g
Surculites g

Taranis g
Tegula g
Tellina p
Terebratulina b
Teredo p
Thesbia g
Thracia p
Thyasira p
Tindaria p
Tivelina p
Tonicella . ch
Tresus p
Trichotropis g
Trigonodesma p
Trochita g
Trochocyathus c
Trochus g
Trophon g
Trophonopsis g
Trophosycon g
Turbonilla g
Turcica g
Turcicula g
Turricula g
Turritella g

Umpquaia g
Urosalpinx g
Uzita p
Velutina g
Venercardia p
Vertipecten p
Volutocorbis g
Whitneyella g
Zenophora g
Yoldia p
Zirfaea p

Siletz, Roseb., Locking., Flourn., Tyee, Yamhill, Elkton, Bateman, Spencer, Cowlitz, Coaledo, Nestucc., Bastend., Keasey, Gries R., Eugene, Tunnel, Pittsb., Alsea, Scapp., Yaquina, Butte Cr., Nye , Astoria, Tarheel, Empire, Elk R., Bandon

137

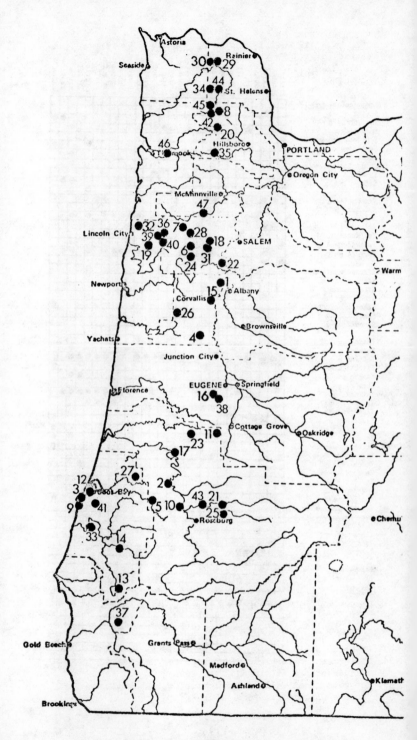

EOCENE FOSSIL INVERTEBRATE LOCALITIES

1. Albany (5 mi. NW of), Benton Co. Spencer Fm. Refs. 232.
2. Basket Point, Douglas Co. Elkton Fm. Refs. 12d, 12e,
   212, 217, 230.
3. Bastendorff Beach, Coos Head, Coos Co. Bastendorff Shale.
   Eocene/Oligocene. Refs. 12d, 12e, 51.
4. Bellfountain, Benton Co. Spencer Fm. Refs. 232.
5. Bottom Creek, Douglas Co. Bateman Fm. Refs. 12d, 12e.
6. Boulder Camp on Rickreall Creek, Polk Co. Yamhill Fm.
   Refs. 21.
7. Buell, Polk Co. Yamhill Fm. Refs. 21, 201d.
8. Buxton (N of on railroad cut), Washington Co. Keasey Fm./
   Pittsburg Bluff Fm./ Scappose Fm. Eocene/Oligocene/Miocene.
   Refs. 142, 229.
9. Cape Arago, Coos Co. Coaledo Fm. Refs. 12d, 12e, 201a, 217.
10. Cleveland, NW of Roseburg, Douglas Co. Refs. 130a, 217.
11. Comstock (5 mi. SW of Cottage Grove), Douglas Co. Elkton Fm.
    Refs. 12d, 12e, 83, 130a, 217.
12. Coos Bay area, Coos Co. Coaledo Fm. Refs. 12d, 12e, 20,
    51, 201a, 217, 232.
13. Coquille River (south fork; Eden Ridge/Powers area), Coos Co.
    Lookingglass Fm./Tyee Fm. Refs. 12d, 12e.
14. Coquille River (middle fork), Coos Co. Roseburg Fm./
    Lookingglass Fm. Refs. 12d, 12e, 83, 130a, 212, 217, 232.
15. Corvallis, Benton Co. Spencer Fm. Refs. 232.
16. Coyote Creek valley, W of Eugene, Lane Co. Spencer Fm.
    Refs. 217, 224.
17. Elkton, Douglas Co. Elkton Fm. Refs. 12d, 12e, 212.
18. Ellendale Basalt Quarry, W of Dallas, Polk Co. Siletz River
    Volcanics. Refs. 12c, 80c, 212, 201d.
19. Euchre Mtn., Lincoln Co. Siletz River Volcanics. Refs. 194.
20. Gales Creek valley, Washington Co. Keasey Fm. Eocene/
    Oligocene. Refs. 180b, 201b.
21. Glide, Douglas Co. Lookingglass Fm. Refs. 12d, 12e, 83,
    130a, 217, 232.
22. Helmick Hill, Polk Co. Spencer Fm. Refs. 12c, 180b.
23. Jack Creek, NW of Drain, Douglas Co. ?Tyee Fm. Refs. 83.
24. Little Luckiamute River, Polk Co. Yamhill Fm. Refs. 12c.
25. Little River, SE of Glide, Douglas Co. Lookingglass Fm.
    Refs. 130a, 155a, 217, 232.
26. Mary's Peak, Benton Co. Siletz River Volcanics.
    Refs. 80c, 212.
27. Matson Creek, Coos Co. Tyee Fm. Refs. 12e.
28. Mill Creek, Polk Co. Yamhill Fm. Refs. 12c, 13, 180b.
29. Mist, Columbia Co. Keasey Fm. Eocene/Oligocene.
    Refs. 142b, 143, 180b, 201b, 229, 247.
30. Nehalem River valley, Columbia Co. Cowlitz Fm./Keasey Fm.
    Eocene/Oligocene. Refs. 201b, 229, 232.
31. Oregon Portland Cement Co., SW of Dallas, Polk Co. Yamhill
    Fm. Refs. 12c, 21, 80c, 180b, 201d.
32. Otis Junction, Lincoln Co. Siletz River Volcanics.
    Refs. 80c, 194.
33. Riverton area, Coos Co. Coaledo Fm. Refs. 12d, 12e, 232.
34. Rock Creek, Columbia Co. Cowlitz Fm./Keasey Fm. Eocene/
    Oligocene. Refs. 180b, 201b, 229.
35. Scoggins Creek valley, Washington Co. ?Yamhill Fm.
    Eocene/Oligocene. Refs. 201b.

EOCENE FOSSIL INVERTEBRATE LOCALITIES

36. Siletz River (north fork), Polk Co. Siletz River Volcanics.
    Refs. 12c.
37. Snout Creek Road, 3 mi. E of Agness, Curry Co. Flournoy Fm.
    Refs. 153.
38. Spencer Creek, W of Eugene, Lane Co. Spencer Fm.
    Refs. 215, 224.
39. Stott Mtn., Lincoln Co. Yamhill Fm. Refs. 12c.
40. Sugarloaf Mtn., Polk Co. Yamhill Fm. Refs. 12c.
41. Summer, Coos Co. Coaledo Fm. Refs. 232.
42. Timber, Washington Co. Cowlitz Fm./Keasey Fm. Eocene/
    Oligocene. Refs. 80b, 180b, 229, 228, 143.
43. Umpqua River (north fork at Lone Rock), Douglas Co.
    Roseburg Fm. Refs. 12d, 12e, 212, 217.
44. Vernonia, Columbia Co. Keasey Fm. Eocene/Oligocene.
    Refs. 180b, 143, 229.
45. Wolf Creek, Washington/Clatsop Co. Keasey Fm./Cowlitz Fm.
    Eocene/Oligocene. Refs. 80b, 143, 220.
46. Wilson River, Tillamook Co. ?Nestucca Fm. Refs. 228.
47. Yamhill River (E of Sheridan), Yamhill Co. Nestucca Fm.
    Refs. 13.

OLIGOCENE FOSSIL INVERTEBRATE LOCALITIES

1. Alsea Bay, Lincoln Co. Alsea Fm./Yaquina Fm. Oligocene/
   Miocene. Refs. 142b, 198, 223.
2. Bastendorff Beach, Coos Head, Coos Co. Bastendorff Shale.
   Eocene/Oligocene. Refs. 12d, 12e, 51.
3. Brownsville (SW of on Calapooya River), Linn Co. Eugene Fm.
   Refs. 77.
4. Butte Creek (Scotts Mills) valley, Clackamas Co. "Butte
   Creek Beds". Oligocene/Miocene. Refs. 53, 76, 152, 159, 23
5. Buxton (N of on railroad cut), Washington Co. Keasey Fm./
   Pittsburg Bluff Fm./Scappose Fm. Eocene/Oligocene/Miocene.
   Refs. 142b, 227.
6. Clatskanie River (headwaters), Columbia Co. Scappose Fm.
   Oligocene/Miocene. Refs. 229.
7. Coal Creek, Columbia Co. Pittsburg Bluff Fm. Refs. 142b.
8. Coburg, Lane Co. Eugene Fm. Refs. 80, 230.
9. Conyers Creek, Columbia Co. ?Gries Ranch Beds. Refs. 12f,
   142b, 228, 229, 237.
10. Devil's Punch Bowl, Lincoln Co. Yaquina Fm. Oligocene/
    Miocene. Refs. 223.
11. Eugene area, Lane Co. Eugene Fm. Refs. 80, 142b, 180b,
    201c, 230.
12. Eola Hills, Polk Co. Eugene Fm. Refs. 12c, 80, 180b,
    201d, 230.
13. Gales Creek valley, Washington Co. Keasey Fm. Eocene/
    Oligocene. Refs. 180b, 201b.
14. Holmes Gap, N of Rickreall, Polk Co. Eugene Fm.
    Refs. 12c, 80, 180b, 201d.
15. Independence (E of), Marion Co. Eugene Fm. Refs. 80.

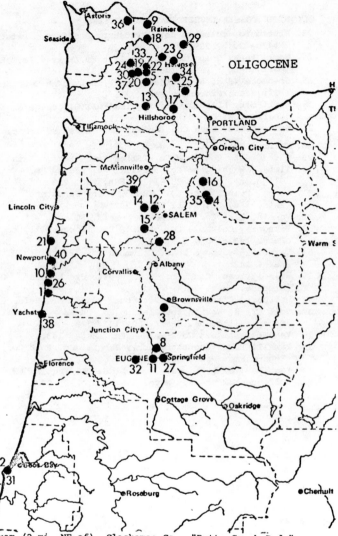

OLIGOCENE

16. Marquam (2 mi. NE of), Clackamas Co. "Butte Creek Beds".
    Oligocene/Miocene. Refs. 76.
17. McKay Creek, Washington Co. Scappose Fm. Oligocene/Miocene.
    Refs. 216.
18. Mist, Columbia Co. Keasey Fm. Eocene/Oligocene.
    Refs. 142b, 143, 180b, 201b, 229, 247.
19. Nehalem River (east fork), Columbia Co. Pittsburg Bluff Fm.
    Refs. 49, 142b, 229, 230, 201b, 143.
20. Nehalem River valley, Columbia Co. Cowlitz Fm./Keasey Fm.
    Eocene/Oligocene. Refs. 201b, 229, 232.
21. Otter Rock, Lincoln Co. Yaquina Fm./Astoria Fm. Oligocene/
    Miocene. Refs. 142, 201, 223.
22. Pebble Creek, Columbia Co. Pittsburg Bluff Fm. Refs. 142b,
    229.

141

OLIGOCENE FOSSIL INVERTEBRATE LOCALITIES
23. Pittsburg, Columbia Co. Pittsburg Bluff Fm. Refs. 49,
    142b, 201b, 229, 230.
24. Rock Creek, Columbia Co. Cowlitz Fm./Keasey Fm. Eocene/
    Oligocene. Refs. 180b, 201b, 229.
25. Scappose (S of on Jackson Creek), Multnomah Co. Scappose Fm.
    Oligocene/Miocene. Refs. 216.
26. Seal Rock, Lincoln Co. Yaquina Fm. Oligocene/Miocene.
    Refs. 223.
27. Springfield, Lane Co. Eugene Fm. Refs. 80, 201c, 230.

28. Talbot, Marion Co. Eugene Fm. Refs. 80.
29. Tide Creek, Columbia Co. ?Gries Ranch Beds. Refs. 238.
30. Timber, Washington Co. Cowlitz Fm./Keasey Fm. Eocene/
    Oligocene. Refs. 80, 143, 180b, 228, 229.
31. Tunnel Point, Coos Co. Tunnel Point Sandstone.
    Refs. 12d, 80, 142b, 180b, 232.
32. Veneta, Lane Co. Eugene Fm. Refs. 80.
33. Vernonia, Columbia Co. Keasey Fm. Eocene/Oligocene.
    Refs. 143, 180b, 229.
34. Vernonia-Scappose road, Columbia Co. Pittsburg Bluff Fm.
    Refs. 142b.
35. Wilhoit road, Clackamas Co. "Butte Creek Beds". Oligocene/
    Miocene. Refs. 76.
36. Westport, Clatsop Co. Pittsburg Bluff Fm. Refs. 142b, 230.
37. Wolf Creek, Washington/Clatsop Co. Keasey Fm./Cowlitz Fm.
    Eocene/Oligocene. Refs. 80, 143, 229.
38. Yachats, Lincoln Co. ?Alsea Fm. Refs. 143.
39. Yamhill River (south branch), Yamhill Co./Polk Co.
    Eugene Fm. Refs. 80.
40. Yaquina (town), Lincoln Co. Alsea Fm./Yaquina Fm.
    Oligocene/Miocene. Refs. 198, 201, 223, 232.

MIOCENE FOSSIL INVERTEBRATE LOCALITIES
 1. Alsea Bay, Lincoln Co. Alsea Fm./Yaquina Fm. Oligocene/
    Miocene. Refs. 142b, 197, 223.
 2. Astoria, Clatsop Co. Astoria Fm. Refs. 1b, 39, 43, 50,
    142, 230.
 3. Bandon, Coos Co. ?Empire Fm. Refs. 211.
 4. Beaver Creek, Tillamook Co. Astoria Fm. Refs. 142.
 5. Big Creek, Clatsop Co. Astoria Fm. Refs. 142.
 6. Butte Creek (Scotts Mills) valley, Clackamas Co. "Butte
    Creek Beds". Oligocene/Miocene. Refs. 53, 76, 152, 159, 2
 7. Buxton (N of on railroad cut), Washington Co. Keasey Fm./
    Pittsburg Bluff Fm./Scappose Fm. Eocene/Oligocene/Miocene.
    Refs. 142b, 229.
 8. Cannon Beach (at mouth of Elk Creek), Clatsop Co. Astoria Fm
    Refs. 142, 230.
 9. Cape Kiwanda, Tillamook Co. Astoria Fm. Refs. 142.
10. Clatskanie River (headwaters), Columbia Co. Scappose Fm.
    Oligocene/Miocene. Refs. 229.
11. Coos Head, Coos Co. Empire Fm. Refs. 7, 12, 38, 43,
    49c, 86, 232.

MIOCENE

MIOCENE FOSSIL INVERTEBRATE LOCALITIES

12. Depoe Bay, Lincoln Co. Astoria Fm. Refs. 44, 142, 201.
13. Devil's Punch Bowl, Lincoln Co. Yaquina Fm. Oligocene/
    Miocene. Refs. 223.
14. Fogarty Creek (Boiler Bay), Lincoln Co. Astoria Fm.
    Refs. 44, 142, 201.
15. Fossil Point, Coos Bay. Coos Co. Empire Fm. Refs. 7, 12,
    12d, 43, 49c, 86, 142, 180b, 211, 232.
16. Marquam (2 mi. NE of), Clackamas Co. "Butte Creek Beds".
    Oligocene/Miocene. Refs. 76.
17. McKay Creek, Washington Co. Scappose Fm. Oligocene/
    Miocene. Refs. 216.
18. Miami River, Tillamook Co. Astoria Fm.
    Refs. 230.
19. Newport, Lincoln Co. Nye Mudstone/Astoria Fm.
    Refs. 142, 142a, 158, 180b, 196, 201, 232.
20. Otter Rock, Lincoln Co. Yaquina Fm./Astoria Fm. Oligocene/
    Miocene. Refs. 142, 201, 223.
21. Pigeon Point, Coos Bay, Coos Co. Tarheel Fm./Empire Fm.
    Refs. 7.
22. Port Orford (SE of on Hubbard Creek), Curry Co. No Fm.
    Refs. 104.
23. Saddle Mtn., Clatsop Co. Astoria Fm. Refs. 142.
24. Scappose (S of on Jackson Creek), Multnomah Co. Scappose Fm.
    Oligocene/Miocene. Refs. 216.
25. Schooner Creek, Lincoln Co. Astoria Fm. Refs. 142.
26. Seal Rock, Lincoln Co. Yaquina Fm. Oligocene/Miocene.
    Refs. 223.
27. Spencer Creek, Lincoln Co. Astoria Fm. Refs. 142,
    158, 201.
28. Tillamook, Tillamook Co. Astoria Fm. Refs. 142, 230.
29. Vernonia-Scappose Road (at Wilark), Columbia Co. Astoria
    Fm. Refs. 142.
30. Wade Creek, Lincoln Co. Astoria Fm. Refs. 142, 158.
31. Wilhoit Road, Clackamas Co. "Butte Creek Beds". Oligocene/
    Miocene. Refs. 76.
32. Yaquina Bay, Lincoln Co. Nye Mudstone. Refs. 142, 158, 196.
33. Yaquina Head, Lincoln Co. Astoria Fm. Refs. 142, 232.
34. Yaquina (town), Lincoln Co. Alsea Fm./Yaquina Fm.
    Oligocene/Miocene. Refs. 196, 201, 223, 232.

PLIOCENE/PLEISTOCENE FOSSIL INVERTEBRATE LOCALITIES
  1. Cape Blanco, Curry Co. Elk River Beds/Port Orford Fm./
       Terrace deposits. Pliocene/Pleistocene.
       Refs. 1, 12a, 35.
  2. Coquille Point, Coos Co. No Fm. Pleistocene. Refs. 247a.
  3. Grave Point, Coos Co. No Fm. Pleistocene. Refs. 247a.
  4. Newport, Lincoln Co. Coquille Fm. Pleistocene. Refs. 12b.

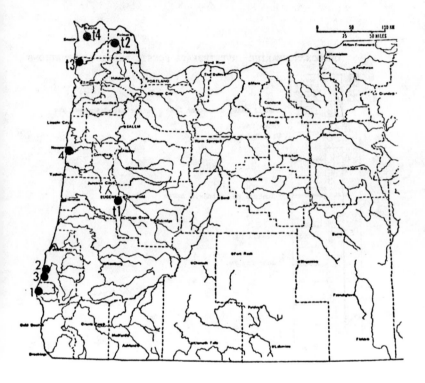

OREGON TRACE FOSSIL LOCALITIES

  Following is a list of Oregon trace fossil localities. The
  reference numbers are to authors listed in the bibliography
  who have published on the specific localities.

  1. Eugene, Lane Co. Eugene Fm. Oligocene. Refs. 80.
  2. Mist, Columbia Co. Keasey Fm. Oligocene. Refs. 2.
  3. Oswald West State Park, Tillamook Co. Oligocene mudstones.
       Oligocene. Refs. 49.
  4. Youngs River, Clatsop Co. Astoria Fm. Miocene. Refs. 63.

145

OREGON NONMARINE INVERTEBRATE FOSSIL BEARING FORMATIONS

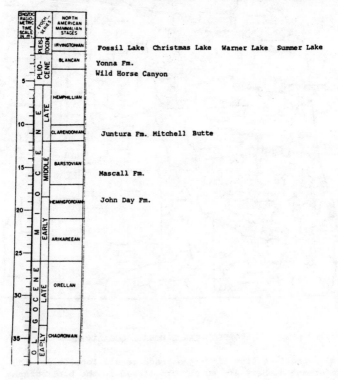

Fossil Lake   Christmas Lake   Warner Lake   Summer Lake

Yonna Fm.

Wild Horse Canyon

Juntura Fm.   Mitchell Butte

Mascall Fm.

John Day Fm.

The following are fossil invertebrate ranges
listed in the published literature.

OREGON FOSSIL FRESHWATER INVERTEBRATE RANGES

Taxon	Fossil Lake	Christmas Lake	Warner Lake	Summer Lake	Yonna Fm.	Wild Horse Canyon	Juntura Fm.	Mitchell Butte	John Day Valley
Ammonitella (gastropoda)									●
Amnicola (gastropoda)				●					
Carinifex (gastropoda)	●						●		
Epiphragmophora (gastropoda)									●
Fluminicola (gastropoda)							●		
Gastrodonta (gastropoda									●
Helicina (gastropoda)									●
Helisoma (gastropoda)	●								
Lanx (gastropoda)			●		●				
Limnophysa (gastropoda)	●								
Lymnea (gastropoda)	●				●				●
Oreohelix (gastropoda)									●
Paludestrina (gastropoda)			●						
Parapholyx (gastropoda)			●						
Physa (gastropoda)					●				
Pisidium (pelecypoda)	●	●				●	●		
Polygyra (gastropoda)									●
Polygyrella (gastropoda)									●
Planorbis (gastropoda)	●	●							
Promenetus (gastropoda)	●								
Pyramidula (gastropoda)									●
Pyrgulopsis (gastropoda)									●
Radix (gastropoda)							●		
Rhiostoma (gastropoda)									●
Sphaerium (pelecypoda)	●				●		●	●	
Unio (pelecypoda)									●
Valvata (gastropoda)	●	●			●				
Viviparus (gastropoda)							●	●	
Vorticifex (gastropoda)			●		●				

147

FRESHWATER FOSSIL INVERTEBRATE LOCALITIES
1. Bear Creek valley, Crook Co. John Day Fm. Miocene.
   Refs. 113.
2. Black Butte, Malheur Co. Juntura Fm. Clarendonian (Miocene).
   Refs. 209a.
3. Fossil Lake-Christmas Lake, Lake Co. No Fm. Pleistocene.
   Refs. 4, 41, 209.
4. John Day River valley, Grant Co. John Day Fm./Mascall Fm.
   Late Oligocene/early Miocene. Refs. 72, 202, 202a.
5. Klamath Falls (15 mi. NE of at Wilson's Quarry Pit),
   Klamath Co. Yonna Fm. Pliocene. Refs. 148.
6. Mitchell Butte, Malheur Co. No Fm. Clarendonian (Miocene).
   Refs. 209a.
7. Summer Lake, Lake Co. No Fm. Pleistocene. Refs. 72a, 72b.
8. Warner Lake, Lake Co. No Fm. Pleistocene. Refs. 72a, 72b.
9. Wild Horse Canyon (W of Lake Owyhee), Malheur Co.
   No Fm. Pliocene. Refs. 209.

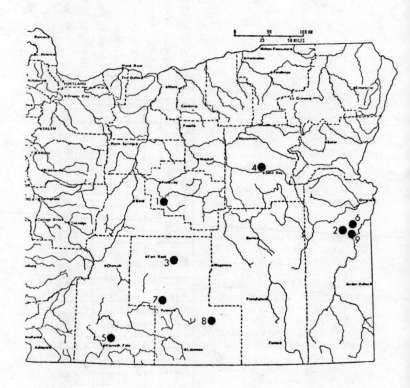

OREGON TRACE FOSSILS

Trace fossils are a group that have assumed increased importance in paleontology in the past few years. A trace fossil is an impression such as a track or burrow of an organism. Because the animal is only rarely found in association with its burrow or tracks, most of the trace fossils are of unknown biological affinity. In some cases the animal is found associated with its burrow. "Ship worm" bored wood is often found with the boring organism, a pelecypod, still intact in the burrow. Hickman (1969) was able to extract almost 50 specimens of the genus *Martesia* from a single piece of fossil wood in the Eugene Formation. Often trace fossils are more common than skeletal or shell material of organisms. Unless it periodically molts, an invertebrate organism ordinarily leaves one fossil specimen. Trace fossils can, on the other hand, be left by the thousands by a single individual as it moves through an environment. In only a few cases are the trace fossils useful for geologic dating. By far the most useful aspect of these fossils is their utility as an environmental indicator. Paleontologists have, for example, learned to readily distinguish the burrows of shallow water crabs and worms from similar but distinctive burrows of outer continental shelf, deep environments.

Although trace fossils are abundant in both marine and non-marine rocks in Oregon, only a few have been treated in the literature. Frey and Cowles (1972) have described and reconstructed tube burrows from the Miocene Astoria Formation and speculate that the burrowing organism, Tisoa, may have been a variety of shrimp. Niem (et al., 1973) reports several trace fossils from an Oligocene mudstone in the field guidebook on northwest Oregon Cenozoic stratigraphy. These trace fossils include burrow patterns such as *Zoophycos* as well as fecal ribbons, *Scalarituba*, and several other types of burrows. Their conclusions for the Oligocene paleoenvironment near Oswald State Park are that the unit was deposited in bathyal or greater depths. Somewhat flattened tubes in the Keasey Formation are described by Adegoke (1967) as probable burrows of marine pogonophoran worms. Today pogonophorans are typically found at bathyal and abyssal depths. Only recently attention was focused on modern pogonophorans living in and around submarine volcanic vents near the Galapagos Islands in water ten to twelve thousand feet deep.

TRACE FOSSILS

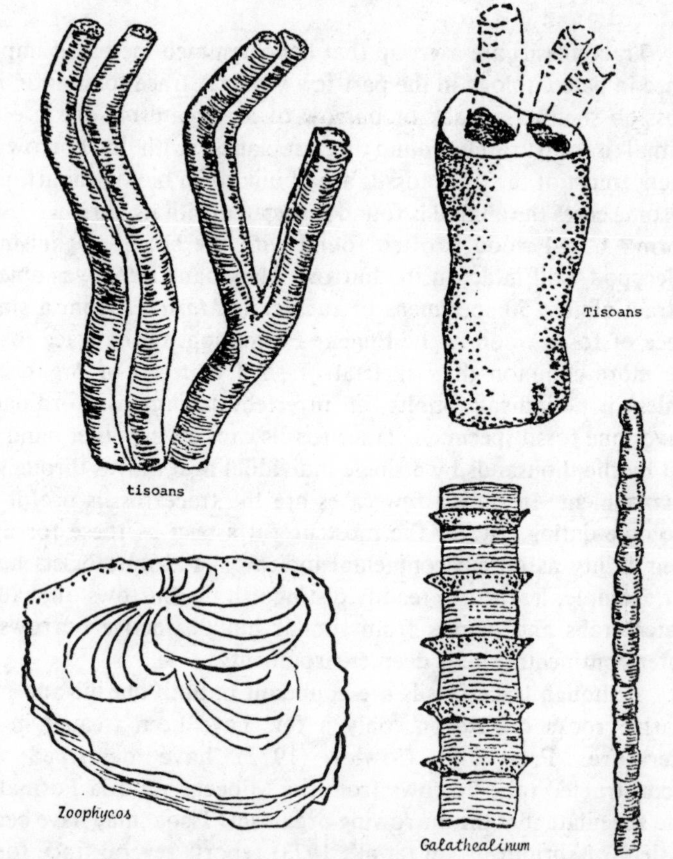

tisoans

Tisoans

Zoophycos

Galathealinum

Lamclёisabella

Martesia

"shipworm" bored fossil wood

1"

OREGON ARTHROPODA/TRACE FOSSIL BEARING FORMATIONS

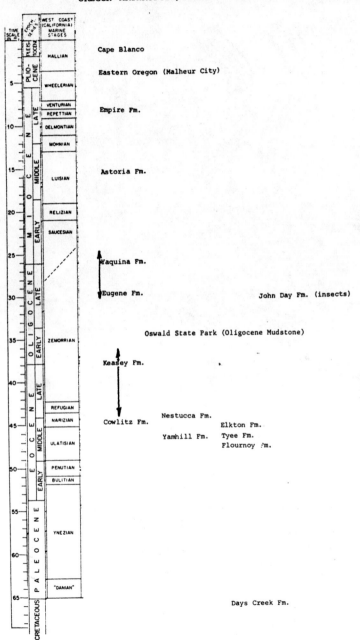

Cape Blanco

Eastern Oregon (Malheur City)

Empire Fm.

Astoria Fm.

Yaquina Fm.

Eugene Fm.   John Day Fm. (insects)

Oswald State Park (Oligocene Mudstone)

Keasey Fm.

Cowlitz Fm.   Nestucca Fm.
                              Elkton Fm.
                 Yamhill Fm.   Tyee Fm.
                              Flournoy Fm.

Days Creek Fm.

OREGON FOSSIL ARTHROPODA

There are many groups of fossil arthropods or "joint-legged" invertebrates, but in the Oregon fossil record three subclasses within the Crustacea are important: the Malacostraca, or crabs and shrimp; the Cirripedia, or barnacles; and the Ostracoda, microfossils similar to a flea in appearance and size with a clam-like shell of paired valves. The ostracod valves are hinged at the top and living forms are able to withdraw the appendages into the valves and snap them tightly shut. At the present time, other than brief mention, there are no published papers on Oregon fossil ostracods. Regardless of this, these organisms are not uncommon in the Oregon fossil record. Because of their size and calcarious composition, ostracods turn up in the extraction process for foraminifera. Both marine and nonmarine ostracods occur in Oregon, but the former are much more common and diverse. In the North American Gulf Coast these fossils are important age and paleoenvironmental indicators to paleontologists in the petroleum industry.

In one of the earliest references to Oregon Crustacea, crabs and barnacles from the Astoria Formation at Astoria were reported by J. Dana of the Wilkes Exploring Expedition. Barnacles are very common in Oregon in association with fossil molluscs. Because of their shallow water habitat, the plates of barnacles are usually scattered and show beach wear. The distinctive grooved triangular plates are easily recognized and make up significant portions of some sediments. Zullo (1969a) has described Pleistocene barnacles found with crabs from the Cape Blanco area. In a separate paper Zullo (1969b) reports from the same locality the occurrence of a tiny pinnotherid crab, *Pinnixa*, fossilized where it lived within the valves of the pelecypod, *Tresus capax*. Symbiotic associations of this type are extremely rare in the fossil record because of post mortem wave action.

Barnacles still attached to fossilized wood have been described from the Oligocene Eugene Formation by Hickman (1969). Within some mollusc bearing levels of the Oligocene/Miocene "Butte Creek Beds" in the Western Cascades, barnacle plates make up over 75% of the invertebrate assemblage (Orr and Faulhaber, 1975). A single fossil isopod has been reported from the Oligocene Keasey Formation near Vernonia in the Newsletter of the Geological Society of Oregon County (1964).

FOSSIL CRABS

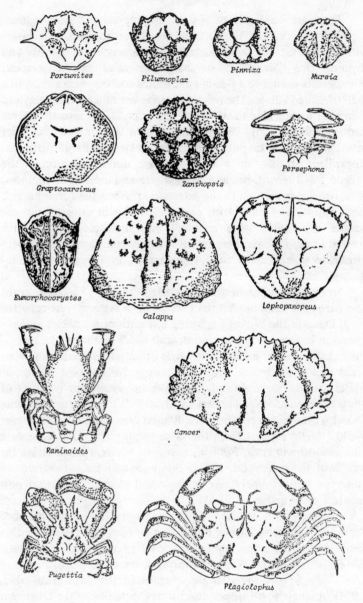

Portunites   Pilumnoplax   Pinnixa   Mursia

Graptocarcinus   Zanthopsis   Persephona

Eumorphocorystes   Calappa   Lophopanopeus

Raninoides   Cancer

Pugettia   Plagiolophus

By far the most common fossil arthropoda in Oregon are the Decapoda, or lobsters, crabs and shrimp. There are many types of arthropods in the coastal ocean today, and these ten-legged Crustacea are certainly one of the most diverse. Crabs and lobsters occupy a large variety of ecologic niches, but most are predatory types or scavengers. Most crabs are vagrant benthonic organisms walking on the ocean bottom, but many are capable of intermittent swimming and may have appendages (tail, legs) modified for this purpose. The wide range of marine habitats and niches yields a considerable array of sizes and shapes in the Decapods.

Arthropods are characterized by molting or "ecdysis" stages as they grow and add segments. Each new growth stage is accompanied by the peeling off of the old shell and the hardening in place of a new shell. Cast off shells or carapaces are often destroyed by wave action or by other organisms, but they may be preserved as fossils. One arthropod specimen, therefore, might be responsible for several discrete fossils. Crabs can also often be differentiated sexually by the shape of the abdominal appendage. The segmented tail typically is thin and pointed on males and broad and rounded on females.

In spite of the fact that fossil crabs are well-preserved and common in Oregon, there are only a few published references to them. Considering briefly the literature on this group, the most definitive work is that of Mary Rathbun (1926). In this monograph she treats Mesozoic and Tertiary fossil "stalk-eyed Crustacea" of the entire Pacific Coast including many from Oregon. In addition to marine lobsters and crabs, Rathbun reports the freshwater crayfish, *Astacus*, from eastern Oregon. In Weaver's monograph (1942) on Oregon and Washington Tertiary invertebrates, he notes several shrimp, barnacles and crabs. A more recent study of modern and fossil crabs of the family Cancridae was made by Dale Nations (1969). He has shown that although this family is found worldwide, only the North American West Coast bears a complete fossil record of the group back to the Miocene. Lobsters about the size of modern crayfish have been described by Feldmann (1974) from the Days Creek Formation (Cretaceous, Hauterivian) in Curry and Douglas Counties in southwest Oregon. Cretaceous Hornbrook exposures on Interstate 5 between Ashland and the southern Oregon border yield abundant smaller shrimp-like forms. These Mesozoic forms are usually poorly preserved, but not rare. Several Eocene crab species are reported in the area of southwest Oregon near Agness from exposures of the Flournoy Formation

ARTHROPODA

Balanus

Balanus

ostracod

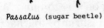

Stomatochlora (dragonfly)

after Cockerell, 1927

Cecidomyia

(gall midge)

Astacus

Passalus (sugar beetle)

Upogebia

Hoploparia

Callianassa

1"

156

(Orr and Kooser, 1971; Kooser and Orr, 1973). Moore (1963) reports the occurrence of the small, shallow-water mud shrimp, *Upogebia*, in the Astoria Formation. This worldwide genus which inhabits rock burrows is known from Jurassic through the Holocene, but the Astoria Formation species may be the first from Miocene rocks.

Most well-preserved crabs occur in sedimentary structures called "concretions." Concretions are discrete accumulations of cementing material in sediments wherein a particle such as a wood fragment, mollusc shell, or other fossil acts as a center of cementation. This process is not unlike the process of making candles by immersing a wick (the fossil) in hot wax (the cement). Cementing material such as calcium carbonate in solution moving slowly through porous rock will accumulate around a crystallization site such as a fossil. Ordinarily concretions are harder than the entombing rock and may often collect as cobbles at the base of an exposure as the rock weathers and crumbles. To extract the fossil, the concretions are split. They are usually oblate in cross-section and may be split along the greatest diameter with sharp blows of a hammer. Fossils preserved in concretion require little further preparation. In some cases a vibra-tool can be used like a dental drill to bring more of the fossil out by removing rock material. Often all of the limbs and even the antennae are preserved in the concretion. As a rule, anywhere fossil molluscs occur in any abundance, fossil crabs are also present.

FOSSIL INSECTS

Insects are a group like snakes, toads and frogs in that their fossil record is only a pale reflection of their abundance and diversity during their lifetime. The best preserved insects in the fossil record are often those captured and entombed in miniature tar pit-like environments. Instead of tar, however, the entrapping medium is simple tree sap. All manner of airborne particles, most commonly pollen, occur in the tree sap along with the insects. With time, a process of slow drying chemically converts the sap to amber. No known occurrences of insects in amber are described in Oregon paleontologic literature, but Cockerell (1927) has described several insects and trace fossils, gall midges, from maple and alder in plant localities. These fossils include wings of dragon flies (Odontata) and jaws of tropical sugar beetles (Coleoptera) from the lower John Day where it crops out on the Gray Ranch near Post. Peterson (1964) has described Trichoptera, caddis fly larval cases, from the John Day Formation in Wheeler County. These consist of fragile organic tubes studded with coarse sand grains.

OREGON FOSSIL ARTHROPODA CLASSIFICATION

The following classification contains Oregon fossil Arthropoda genera listed in the published literature. Taxonomic revision has not been attempted.

CLASS: Crustacea
  SUBCLASS: Ostracoda
  SUBCLASS: Cirripedia
    ORDER: Thoracica (barnacle)
      GENERA: *Balanus*
  SUBCLASS: Malacostraca
    ORDER: Isopoda
    ORDER: Decapoda (10-footed crustaceans, crayfish, lobsters, true crabs)
      GENERA: *Astacus, Calappa, Callianassa, Cancer, Eucrate, Eumorphocorystes, Graptocarcinus, Haplogaster, Hoploparia, Lophopanopeus, Mursia, Palehomola, Persephona, Phyllolithodes, Pilumnoplax, Pinnixa, Plagiolophus, Portunites, Pugettia, Raninoides, Upogebia, Zanthopis.*

CLASS: Insecta
    ORDER: Odonata (dragonfly)
      GENERA: *Somatochlora*
    ORDER: Coleoptera (sugar beetle)
      GENERA: *Passalus*
    ORDER: Diptera (gall midge)
      GENERA: *Cecidomya*
    ORDER: Trichoptera (caddis fly)

The following are fossil arthropoda ranges listed in the published literature.

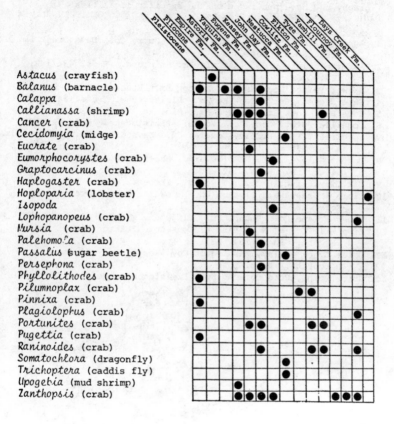

OREGON FOSSIL ARTHROPODA LOCALITIES

Following is a list of Oregon fossil arthropoda localities.
The reference numbers are to authors listed in the bibliography who have published on the specific localities.

1. Agness (1 mi. S of on Rogue River), Curry Co. Days Creek Fm.
   Cretaceous. Refs. 59.
2. Astoria, Clatsop Co. Astoria Fm. Miocene. Refs. 43, 142,
   167, 236.
3. Basket Point, Douglas Co. Elkton Fm. Eocene. Refs. 167, 232.
4. Cape Blanco, Curry Co. No Fm. Pleistocene. Refs. 247b.
5. Coos Bay, Coos Co. Empire Fm. Miocene. Refs. 43, 232.
6. Coquille Point, Coos Co. No Fm. Pleistocene. Refs. 247b.
7. Cow Creek, S of Riddle, Douglas Co. Days Creek Fm. Cretaceous.
   Refs. 59.
8. Dugout Gulch, 2 mi. NE of Clarno, Wheeler Co. John Day Fm.
   Oligocene. Refs. 162.
9. Elk City, Lincoln Co. ?Tyee Fm. ?Eocene. Refs. 167.
10. Elsie (Mishawaka), Clatsop Co. ?Cowlitz Fm. ?Eocene.
    Refs. 167.
11. Eugene, Lane Co. Eugene Fm. Oligocene. Refs. 80, 167, 232.
12. Five Mile Creek, N of Bandon, Coos Co. ?Elkton Fm.
    ?Eocene. Refs. 167, 232.
13. Gales Creek valley, Washington Co. Keasey Fm. Eocene/
    Oligocene. Refs. 201b.
14. Grave Point, Coos Co. No Fm. Pleistocene. Refs. 247b.

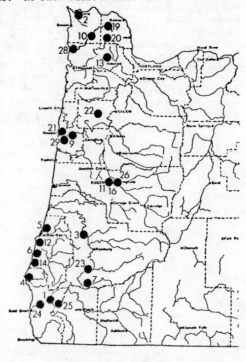

160

15. Gray Ranch (McCullough Ranch) 10 mi. E of Post, Crook Co.
      John Day Fm. Oligocene. Refs. 36.
16. Judkins Point, Lane Co. Eugene Fm. Oligocene. Refs. 167,
      232.
17. Little Valley (5 mi. SE of), Malheur Co. No Fm. ?Miocene
      (freshwater). Refs. 167.
18. Malheur City (Eldorado), Malheur Co. No Fm. ?Pliocene.
      (freshwater). Refs. 167.
19. Mist, Columbia Co. Keasey Fm. Eocene/Oligocene. Refs. 247.
20. Nehalem River valley, Columbia Co. Cowlitz Fm. Eocene.
      Refs. 229.
21. Newport, Lincoln Co. ?Yaquina Fm. Oligocene/Miocene.
      Refs. 167.
22. Oregon Portland Cement Co., SW of Dallas. Polk Co. Yamhill
      Fm. Eocene. Refs. 12c.
23. Riddle, Curry Co. Days Creek Fm. Cretaceous. Refs. 59.
24. Rogue River valley, Curry Co. Days Creek Fm. Cretaceous.
      Refs. 59.
25. Snout Creek Road, 3 mi. E of Agness, Curry Co. Flournoy Fm.
      Eocene. Refs. 106, 153.
26. Springfield (on S Pacific railroad cut), Lane Co. Eugene Fm.
      Oligocene. Refs. 80, 167, 232.
27. Vale, Malheur Co. No Fm. ?Miocene (freshwater). Refs. 167.
28. Wheeler (on S Pacific railroad cut), Tillamook Co.
      ?Nestucca Fm. Eocene. Refs. 167.
29. Yaquina (town), Lincoln Co. ?Alsea Fm./?Yaquina Fm. Oligocene/
      Miocene. Refs. 167, 232.

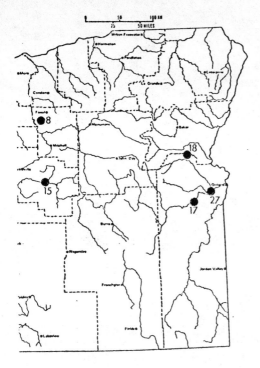

OREGON FOSSIL FRESHWATER FISH

Freshwater fish fossils were first described from Oregon Great Basin localities by Cope. Some of the better known sites include Pleistocene forms from Lake County, Pliocene forms from Klamath County, Miocene from Malheur County, Oligocene and Eocene forms from Wheeler County.

Fossil fish in Oregon are only very rarely preserved in an articulated condition, and the commonest remains are scales and bones. A variety of Eocene freshwater fish are known from the Clarno Formation where it is exposed in roadcuts in western Wheeler County in the Ochoco Pass area. This locality is important for its diversity of fish remains and its stratigraphic position between the famous middle Eocene Green River fish-bearing shales of Wyoming and several other mid-Tertiary localities in western North America. Fish varieties in the Clarno include, among other forms, bowfins, mooneyes, catfish, and suckers. Cavender (1968) has interpreted the locality as a lacustrine (lake) deposit. Bowfin scales from the Ochoco Pass locality are distinctive thick elongate elements and are usually preserved intact. In evolutionary terms the bowfins are a particularly conservative group of fish having changed very little in 70 million years from their first appearance in the Cretaceous. Most of the fish remains at the Eocene Ochoco Pass locality are scales, but bone fragments of vertebra and skull elements are not uncommon. Catfish from Ochoco Pass are reviewed by Lundberg (1975) who also notes a Miocene catfish recovered by Shotwell from the Chalk Butte Formation of southeastern Oregon.

Cavender (1969) has described in some detail articulated specimens of an Oligocene mudminnow, *Novumbra oregonensis*, Cavender, from the Bridge Creek locality of the John Day Formation. These small fish are around 4 inches in length and are preserved in what was a mud substrate of volcanic ash and organic debris. Living mudminnows are characterized in their habitat as living in areas of quiet water with a very fine grain clastic (mud) bottom and dense aquatic vegetation. Mudminnows display a considerable tolerance for temperature and dissolved oxygen variation but somewhat limited tolerance for salinity and current strength variation. The environment suggested for the Bridge Creek locality by these small fish is consistent with that indicated by the mammalian and plant remains from the same unit.

FOSSIL FRESHWATER FISH

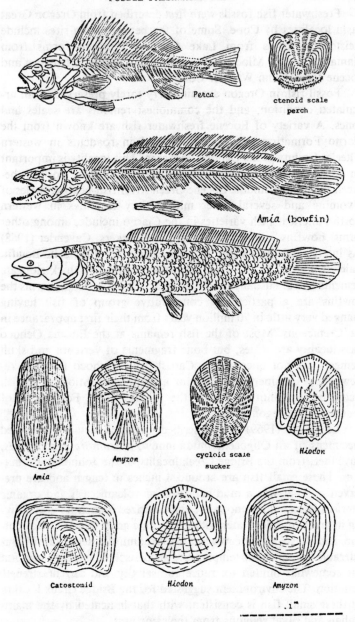

Perca

ctenoid scale
perch

Amia (bowfin)

Amia

Amyzon

cycloid scale
sucker

Hiodon

Catostomid

Hiodon

Amyzon

.1"

164

Elsewhere in eastern Oregon most fossil leaf-bearing lacustrine (lake) sediments of Oligocene and Eocene age contain undescribed scales and bones of freshwater fish. Chaney (1927) records a form from the Oligocene, *Pholidophorus*, found with leaf deposits on the Gray Ranch. This was found with one other bony-plate type of fish not completely identified. Romer elsewhere regards *Pholidophorus* as a Jurassic genus.

Cope (1883c) reported a Miocene locality in Baker County on Willow Creek with several species of fish, crayfish, molluscs and mammals. Several of the fossil fish reported from this locality are not found today living west of the continental divide and are unreported elsewhere in Oregon. Miocene freshwater fish from the Deer Butte Formation of southeastern Oregon are described by Kimmel (1975). This diverse fauna includes salmon, catfish, sunfish, sculpins and whitefish. In addition to large predaceous types such as the genus *Ptychocheilus*, the fauna here includes several mollusc eating fish of the genera *Mylocheilus* and *Idadon*. Jaw elements of these fish contain blunt or rounded mollusc crushing teeth which at first glance are remarkably like those of a primate. Kimmel has been able to build a paleoenvironmental picture of the Deer Butte site by careful geological and biological observations. He notes, for example, that the massive sand and siltstone interbedded with pure volcanic ash at this site suggests either a slow deposition or a rapid deposition with very intermittent volcanic activity. The presence of articulated fish fossils in the siltstones supports the rapid depositional hypothesis as do the conglomerate beds present. A rapid depositional site would be typical of a flood plain or lake edge environment. In the Deer Butte the size of some of the predatory fish, which include three species close to three feet in length, further suggests a large volume habitat such as a lake, but genera typical of a lacustrine (lake) environment are curiously missing. Kimmel's hypotheses to explain these apparent anomalies centers around a very large lacustrine structure capable of including several subenvironments with streams feeding the lake.

Fossil salmon have been recovered from localities of various ages from Miocene to Recent in eastern Oregon. Probably the most exceptional is the late Tertiary Miocene and Pliocene "sabre tooth salmon," *Smilodonichthys rastrosus*, described by Cavender and Miller (1972) from Oregon localities near Gateway and Worden, as well as from other small occurrences along the Columbia River. This species is known only from a few vertebra and a complete skull of 17 inches in length. Skull fragments of even larger

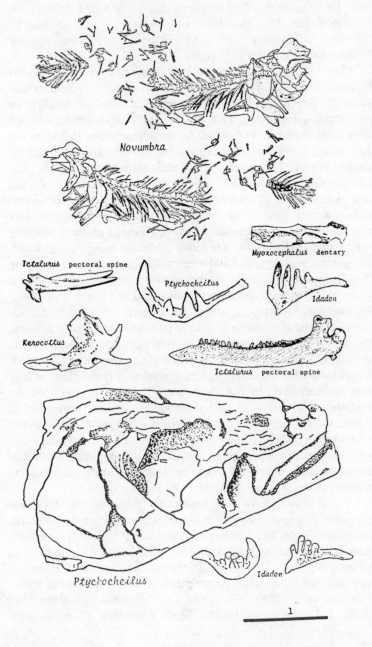

Novumbra

Myoxocephalus dentary

Ictalurus pectoral spine

Ptychocheilus

Idadon

Kerocottus

Ictalurus pectoral spine

Ptychocheilus

Idadon

1

specimens have been recovered and the entire fish may have been as much as eight feet in length. The size of this fish as well as details of its bone structure indicate that, like modern salmon, it was anadromous, splitting its life cycle between freshwater and the ocean. In addition to its massive size, there is evidence that the specimen was a breeding male and as such sported fang-like fighting teeth roughly 1.5 inches in length. In spite of its fierce appearance and name, this great fish probably fed on plankton aided by a delicate and elaborate system of over 100 gill rakers functioning as a sieve.

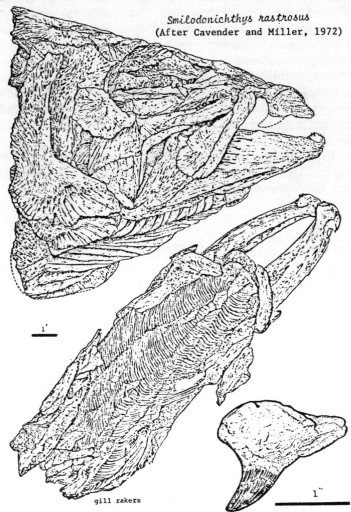

Smilodonichthys rastrosus
(After Cavender and Miller, 1972)

gill rakers

1'

1"

A famous Pleistocene locality for fossil fish in Oregon is in northern Lake County at Fossil Lake where Cope (1883d) described several freshwater taxa including minnows and suckers. Jordan's (1907) revised list of species from this locality included salmon. At this site terrestrial vertebrate (mammalian) remains are found, but fish are by far the most common fossil. The fish bones at Fossil Lake litter the fine sands of what was a large lake during the late Tertiary and Pleistocene. The lack of fine-grained clays in much of the old lake surface deposits has precluded the preservation of fragile scales here, but skull bones and vertebra abound and are easily collected from the loose sand. The shallow environment of the lake has thoroughly scattered the individual bones of fish, and no articulated remains are reported. Allison (1966) interprets the salmon bones in Fossil Lake as an indication of an overflow stage of former Fort Rock Lake, a much larger water body encompassing the present day areas of Fossil Lake, Christmas Lake, and Silver Lake, through an outlet to the Pacific Ocean. Salmon apparently remained in the lake until the end of its existence as a single body of water.

In a summary of late Cenozoic freshwater fish of North America, Uyeno and Miller (1963) revised the identifications of many early reports of fish from sites in eastern Oregon.

OREGON FOSSIL FRESHWATER FISH CLASSIFICATION

The following classification contains Oregon fossil freshwater fish genera listed in the published literature.
Taxonomic revision has not been attempted.

INFRACLASS: Teleostei
  ORDER: Salmoniformes
      FAMILY: Salmonidae (salmon-like)   *Salmon*
         GENERA: *Oncorhynchus, Paleolox, Prosopium, Rhabdofario,*
                 Smilodonichthys.
      FAMILY: Esocidae (salmon-like)
         GENERA: *Esox.*
      FAMILY: Umbridae (mudminnow)   *Mud Minnow*
         GENERA: *Novumbra*
  ORDER: Cypriniformes
      FAMILY: Catostomidae (suckers)
         GENERA: *Amyzon, Catostomus, Chasmites.* *Sucker*
      FAMILY: Cyprinidae (small minnows, dace, shiners)
         GENERA: *Acrocehelius, Alburnops, Anchybopsis, Gila*
                (Siphateles), Idadon, Mylocheilus, Notropis,
                Orthodon, Ptychocheilus.
      FAMILY: Cobitidae
  ORDER: Siluriformes
      FAMILY: Ictaluridae (catfish)
         GENERA: *Ictalurus.*   *Dace*
      FAMILY: Siluridae
  ORDER: Osteoglossiformes   *cat*
      FAMILY: Hiodontidae (mooneyes)
         GENERA: *Hiodon.*
  ORDER: Perciformes   *Mooneye*
      FAMILY: Centrarchidae (sunfish)
         GENERA: *Archoplites.*
      FAMILY: Percidae (perch)
         GENERA: *?Plioplarchus.*
  ORDER: Scorpaeniformes
      FAMILY: Cottidae (sculpin)   *Perch*
         GENERA: *Cottus, Kerocottus, Myoxocephalus.*
INFRACLASS: Holostei
  ORDER: Amiformes
      FAMILY: Amiidae (bowfin)
         GENERA: *Amia.*
                  Sculpin
  ORDER: Pholidophoriformes
      (Halecostomi)

      FAMILY: Pholidophoridae
         GENERA: *Pholidophorus.*   *Bowfin*

OREGON FOSSIL FRESHWATER FISH RANGES

The following are fossil freshwater fish ranges listed in the published literature.

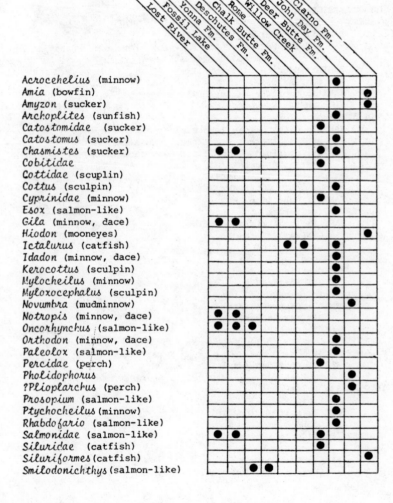

	Lost River	Fossil Lake	Yonna Fm.	Deschutes Fm.	Chalk Butte Fm.	Rome	Willow Creek	Deer Butte Fm.	John Day Fm.	Clarno Fm.
Acrocehelius (minnow)									●	
Amia (bowfin)										●
Amyzon (sucker)										●
Archoplites (sunfish)									●	
Catostomidae (sucker)								●		
Catostomus (sucker)									●	
Chasmistes (sucker)	●	●						●	●	
Cobitidae								●		
Cottidae (scuplin)										
Cottus (sculpin)									●	
Cyprinidae (minnow)								●		
Esox (salmon-like)									●	
Gila (minnow, dace)	●	●								
Hiodon (mooneyes)										●
Ictalurus (catfish)						●	●		●	
Idadon (minnow, dace)									●	
Kerocottus (sculpin)									●	
Mylocheilus (minnow)									●	
Myloxocephalus (sculpin)									●	
Novumbra (mudminnow)									●	
Notropis (minnow, dace)	●	●								
Oncorhynchus (salmon-like)	●	●	●							
Orthodon (minnow, dace)									●	
Paleolox (salmon-like)									●	
Percidae (perch)								●		
Pholidophorus									●	
?Plioplarchus (perch)									●	
Prosopium (salmon-like)									●	
Ptychocheilus (minnow)									●	
Rhabdofario (salmon-like)									●	
Salmonidae (salmon-like)	●	●						●		
Siluridae (catfish)								●		
Siluriformes (catfish)										●
Smilodonichthys (salmon-like)				●	●					

OREGON FOSSIL FISH AND BIRD BEARING FORMATIONS

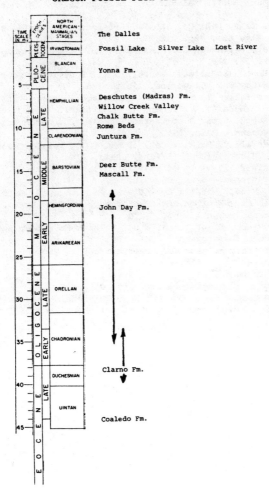

The Dalles

Fossil Lake   Silver Lake   Lost River

Yonna Fm.

Deschutes (Madras) Fm.
Willow Creek Valley
Chalk Butte Fm.
Rome Beds
Juntura Fm.

Deer Butte Fm.
Mascall Fm.

John Day Fm.

Clarno Fm.

Coaledo Fm.

OREGON FOSSIL FRESHWATER FISH LOCALITIES

Following is a list of Oregon fossil freshwater fish localities
The reference numbers are to authors listed in the bibliography
who have published on the specific localities.

1. Allen Ranch, 9 mi. NW of Mitchell (became Wade Ranch), Wheeler
   Co. John Day Fm. Oligocene. Refs. 29a.
2. Arlington, Gilliam Co. No Fm. ?Miocene. Refs. 31.
3. Blackjack Butte, Malheur Co. Deer Butte Fm. Miocene
   (Barstovian). Refs. 99.
4. Fossil Lake, Lake Co. No Fm. Pleistocene. Refs. 4, 31,
   41, 41g, 41m, 78c, 218.
5. Gateway, Jefferson Co. Deschutes Fm. Miocene (Hemphillian).
   Refs. 31, 218.
6. Gray Ranch (McCullough Ranch) 10 mi. E of Post, Crook Co.
   John Day Fm. Oligocene. Refs. 32d.
7. Klamath Falls (15 mi. NE of, at Wilson's Quarry Pit), Klamath
   Co. ?Yonna Fm. Pliocene (Hemphillian). Refs. 148.
8. Knox Ranch, 6 mi. E of Clarno, Wheeler Co. John Day Fm.
   Oligocene. Refs. 29a.
9. Little Valley, Malheur Co., 10 mi. SW of Vale. Chalk Butte
   Fm. Miocene (Hemphillian). Refs. 116.
10. Lost River, 15 mi. E of Klamath Falls. Klamath Co. No Fm.
    Pleistocene. Refs. 96, 218.
11. Ochoco Mtns., Wheeler Co. Clarno Fm. Eocene. Refs. 29, 116.
12. Rome (5 mi. SW of on Dry Creek), Malheur Co. No Fm.
    Miocene (Hemphillian). Refs. 116.
13. Shenk Ranch, SE of Tunnel Mtn., Malheur Co. Deer Butte Fm.
    Miocene (Barstovian). Refs. 99.
14. Tunnel Mtn., Malheur Co. Deer Butte Fm. Miocene (Barstovian).
    Refs. 99.
15. Van Horn Ranch, 12 mi. E of Dayville, Grant Co. John Day Fm.
    Miocene (Hemingfordian). Refs. 41L, 218.
16. Willow Creek, Baker Co. No Fm. ?Miocene. Refs. 41f, 218.
17. Worden, Klamath Co. ?Yonna Fm. Pliocene (Hemphillian).
    Refs. 31.

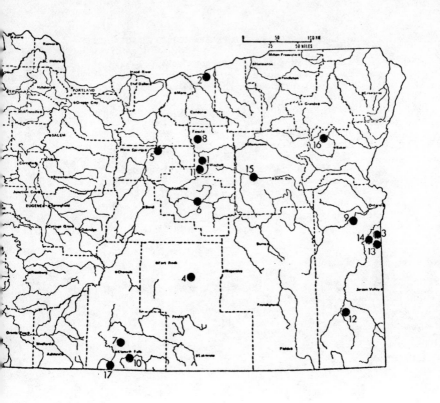

OREGON FOSSIL BIRDS

Ordinarily birds have a comparatively poor fossil record due to their fragile air frame of hollow bones. Very recently in the paleontologic literature a considerable amount of attention has been focused on birds and avian paleontology because of the theory advanced by some paleontologists that birds may be the direct descendants of dinosaurs. Indeed, the presence of feathers is the primary distinction birds possess over reptiles. When one considers the unlikely probability of feathers being preserved as fossils, it is clear that the ancestry of birds will be difficult to trace at best. Flight ability is certainly not unique to birds. Insects, mammals and reptiles have also acquired this skill to varying degrees of mastery.

The large numbers of late Tertiary and Quaternary birds associated with prehistoric inland lakes of eastern Oregon have placed them in the fossil record regardless. The oldest bird remains in the state are of an Eocene auklet described by Miller (1931) from near Coos Bay. The fossil material consists of a leg bone only of the species *Hydrotherikornis oregonus*.

As with fossil fish, the Oligocene/Miocene John Day Formation has yielded little or nothing in the way of fossil birds despite the richness of mammalian faunas from that formation. This doubtless reflects the depositional environment. Condon's interpretation of the depositional environment of the John Day Formation was a lacustrine or a lake environment. Elsewhere, however, in the lacustrine late Tertiary Fossil Lake area we see extremely rich fish and bird faunas. The aeolian or wind-blown dune deposit origin we presently view for the John Day Formation may better explain this notable lack of fish and birds in the Oligocene here. A middle Miocene (Barstovian) *Phasianus* (pheasant) was reported from Paulina Creek (Shufeldt, 1915), and a late Miocene bird collection from near Willow Creek in Baker County included the genera *Larus* (gull), *Limicolavis* (plover), and a *Phalacrocorax* (cormorant) (Shufeldt, 1915). Late Tertiary fossil bird material has been described by Brodkorb (1958, 1961) and Miller (1944) from various lakes now dry in eastern Oregon. These lake areas include the Juntura Basin and Dry Creek area southwest of Rome in Malheur County and the McKay Reservoir near Arlington. Bird fossils in these areas include a stork, flamingo, ducks, cormorant, swan, teal, pheasant and coots. Associated with the avian fauna in these areas is a large mammalian fauna which includes many aquatic

FOSSIL BIRDS

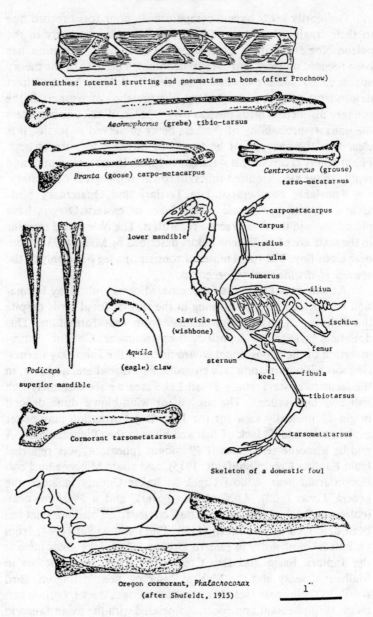

Neornithes: internal strutting and pneumatism in bone (after Prochnow)

Aechmophorus (grebe) tibio-tarsus

Branta (goose) carpo-metacarpus

Centrocercus (grouse) tarso-metatarsus

lower mandible

-carpometacarpus
-carpus
-radius
-ulna
-humerus
-ilium
-ischium

Podiceps superior mandible

Aquila (eagle) claw

clavicle (wishbone)
sternum
keel
-femur
-fibula
-tibiotarsus
-tarsometatarsus

Cormorant tarsometatarsus

Skeleton of a domestic fowl

Oregon cormorant, *Phalacrocorax* (after Shufeldt, 1915)

1"

176

forms such as beaver and otter suggesting an ancient lake close by marshy areas. Although Miller and Brodkorb regard the Dry Creek, McKay Reservoir and Juntura localities as Pliocene in age, mammals from the sites indicate a Hemphillian stage (late Miocene) age.

The greatest volume of fossil bird material has been recovered from Pleistocene sites in the State primarily in dry lakes east of the Cascades. The great diversity of birds at these fossil sites indicates that, as with Klamath Lake today, they were way stations on flyways for migratory birds. It is interesting to note that today, although one half of all birds belong to the order Passeriformes (perching birds), only four of the more than sixty fossil genera in the State are of this Order. This reflects the habitat of the bird groups because more than three-quarters of the Fossil Lake genera are waterbirds or shorebirds, preserved in what amounted to an optimum area for burial and fossilization. Conditions here in the lake environment with soft organic-rich mud and sand gently covering carcasses of dead organisms with only limited wave activity make the fossil record rich despite the inherent fragile nature of the bird skeletons. The collecting sites themselves defy imagination with gently drifting sands today uncovering thousands of the unbroken bird bones in association with remains of freshwater fish and molluscs. Although fish material far outnumbers bird fossils, the careful observer working on hands and knees can quickly make fossil finds without difficulty. Unfortunately none of the skeletons are articulated due to the slow sedimentation rate that characterized these inland lakes. Howard (1946) in reviewing the Fossil Lake site notes the confusion in the literature where the terms Christmas Lake, Fossil Lake, and Silver Lake are often used interchangeably for localities. All these geographic areas at one time may have been parts of the same Pleistocene lake, but study of the various locality data suggests the authors were all referring to one specific site. Howard has made a case for referring to all these sites as "Silver Lake"; however, in view of the present familiarity with the term Fossil Lake, the latter designation is used here.

Condon made the first collections in the Fossil Lake area in 1877. Cope (1878), Miller (1912), Shufeldt (1891, 1912, 1913), Howard (1946, 1964), and Jehl (1967) have all described birds from thousands of bones recovered from this site. A collection of nearly 600 specimens comprising 68 species was assembled by I. Allison between 1939 and 1941 (Allison, 1966). Sixteen of Allison's species are extinct. Birds represented are similar to those at the huge inland lakes of California. The many juvenile specimens found confirm that the birds were nesting at Fossil Lake. Howard regards the Fossil Lake bird locality as correlative with the Rancho LaBrea site in Los Angeles.

Miller (1912) discusses at some length extinction in birds in the late Tertiary. He suggests that while man probably did not directly cause bird extinction, the demise of large mammals certainly contributed to the extinction of large carrion eating birds. One such bird is the huge *Teratornis* of California. Three times as large as a condor, it must have experienced even more difficulty than condors getting airborne after a large meal including a long take off run and ensuing struggle to get air speed. There is, Miller notes, evidence of competition with old world forms, but predators such as cats and mustelids may have contributed to the demise of many bird species. Disease is also suggested as an extinction cause as well as diminished rainfall and the thinning out of forests.

Although not technically "fossils" in the strictest sense, it is interesting to note bird bones described by Miller (1957) in association with early Indian sites in Oregon near The Dalles on the Columbia River. Sites recorded by the above author are around 8,000 years old and reportedly included a vulture (*Coragyps*), condor (*Gymnogyps*), gull (*Larus*) and cormorant (*Phalacrorax*).

OREGON FOSSIL BIRD CLASSIFICATION

The following classification contains Oregon fossil bird genera listed in the published literature. Taxonomic revision has not been attempted.

ORDER

Podicipediformes
(Colymbiformes)
    FAMILY: Podicipedidae (grebe)
        GENERA: *Aechmophorus, Podiceps, Podilymbus.*
Pelicaniformes
    FAMILY: Pelicanidae (pelican)
        GENERA: *Pelecanus.*
    FAMILY: Phalacrocoracidae (cormorant)
        GENERA: *Phalacrocorax*

Ciconiiformes
    FAMILY: Phoenicopteriidae (flamingo)
        GENERA: *Phoenicopterus*
    FAMILY: Palaelodidae
        GENERA: *Megapaloelodus*
    FAMILY: Ardeidae (heron, bittern)
        GENERA: *Ardea, Botaurus*
Anseriformes
    FAMILY: Anatidae (goose, duck, swan, mallard)
        GENERA: *Anabernicula, Anas, Anser, Branta, Charitonetta, Chen, Clangula, Cygnus, Eremochen, Erismatura, Mareca, Melanitta, Mergus, Nettion, Nyroca, Ocyplonessa, Oxyura, Querquedula, Spatula, Sthenelides.*
Falconiformes
    FAMILY: Vulturidae (vulture, condor)
        GENERA: *Coragyps, Gymnogyps.*
    FAMILY: Accipitridae (eagle, hawk)
        GENERA: *Aquila, Circus, Haliaeetus, Hypomorphnus, Neophrontops, Spizaetus.*
    FAMILY: Falconidae (falcon)
        GENERA: *Falco.*
Galliformes
    FAMILY: Phasianidae (grouse, pheasant)
        GENERA: *Centrocercus, Dendragapus, Lophortyx, Palaeotetrix, Phasianus, Pediocetes.*
Ralliformes
    FAMILY: Rallidae (coot, rail)
        GENERA: *Fulica, Rallus.*
Charadriiformes
    FAMILY: Scolopacidae (plover, surfbird, sandpiper, curlew)
        GENERA: *Bartramia, Erolia, Limicolavis, Limnodromus, Lobipes, Numenius, Totanus.*
    FAMILY: Laridae (gull, black tern)
        GENERA: *Larus, Sterna, Childonias.*
    FAMILY: Stercorariidae (jaeger)
        GENERA: *Stercorarius*
    FAMILY: Alcidae (auk)
        GENERA: *Hydrotherikornis.*
    FAMILY: Recurvirostridae (avocet)
        GENERA: *Himantopus, Recurvirostra.*
Strigiformes
    FAMILY: Strigidae (owl)
        GENERA: *Bubo.*
Passeriformes
    FAMILY: Corvidae (raven)
        GENERA: *Corvus.*
    FAMILY: Icteridae (blackbird)
        GENERA: *Agelaius, Euphagus, Sturnella.*
Piciformes
    FAMILY: Picidae (flicker)
        GENERA: *Colpates.*

OREGON FOSSIL

BIRD RANGES

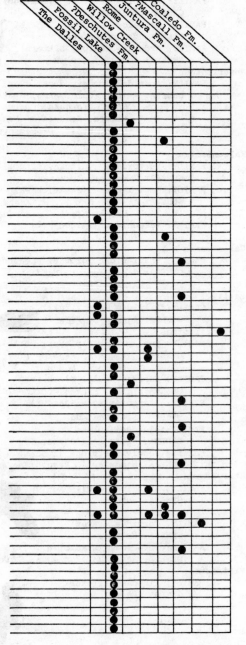

Aechmophorus (grebe)
Agelaius (blackbird)
Anabernicula (pigmy goose)
Anas (duck)
Anser (goose)
Aquila (eagle)
Ardea (heron)
Bartramia (surfbird)
Botaurus (bittern)
Branta (goose)
Bubo (owl)
Centrocercus (grouse)
Charitonetta (duck)
Chen (goose)
Childonias (tern)
Circus (hawk)
Clangula (duck)
Colpates (flicker)
Coragyps (vulture)
Corvus (raven)
Cygnus (swan)
Dendragapus (grouse)
Erismatura (duck)
Eremochen (duck)
Erolia (sandpiper)
Euphagus (blackbird)
Falco (falcon)
Fulica (coot)
Gymnogyps (condor)
Haliaeetus (eagle)
Himantopus (avocet)
Hydrotherikornis (auk)
Hypomorphnus (eagle)
Larus (gull)
Limicolavis (surfbird)
Limnodromus (surfbird)
Lobipes (surfbird)
Lophortyx (pheasant)
Mareca (duck)
Megapaleolodus
Melanitta (surfbird)
Mergus (duck)
Neophrontops (eagle)
Nettion (duck)
Numenius (curlew)
Nyroca (duck)
Ocyplonessa (duck)
Oxyura (duck)
Palaeotetrix (grouse)
Pediocetes (grouse)
Pelecanus (pelican)
Phalacrocorax (cormorant)
Phasianus (pheasant)
Phoenicopterus (flamingo)
Podiceps (grebe)
Podilymbus (grebe)
Querquedula (duck)
Rallus (rail)
Recurvirostra (avocet)
Spatula (duck)
Spizaetus (eagle)
Stercorarius (jaeger)
Sterna (gull)
Sthenelides (swan)
Sturnella (blackbird)
Totanus (shorebird)

OREGON FOSSIL FISH AND BIRD BEARING FORMATIONS

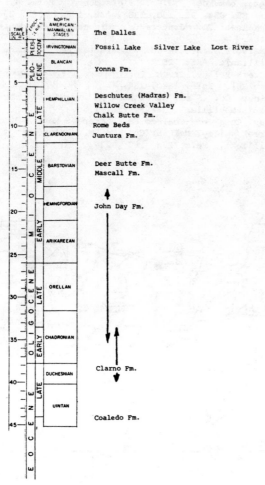

TIME SCALE (M.Y.)	EPOCH	SERIES	NORTH AMERICAN MAMMALIAN STAGES	
	PLEIS- TOCENE		IRVINGTONIAN	The Dalles
	PLIO- CENE			Fossil Lake Silver Lake Lost River
			BLANCAN	Yonna Fm.
5				
		LATE	HEMPHILLIAN	Deschutes (Madras) Fm. Willow Creek Valley Chalk Butte Fm.
10			CLARENDONIAN	Rome Beds Juntura Fm.
15	MIOCENE	MIDDLE	BARSTOVIAN	Deer Butte Fm. Mascall Fm.
20			HEMINGFORDIAN	John Day Fm.
25		EARLY	ARIKAREEAN	
30		LATE	ORELLAN	
35	OLIGOCENE	EARLY	CHADRONIAN	
40		LATE	DUCHESNIAN	Clarno Fm.
45	EOCENE		UINTAN	Coaledo Fm.

181

OREGON FOSSIL BIRD LOCALITIES

Following is a list of Oregon fossil bird localities. The
reference numbers are to authors listed in the bibliography
who have published on the specific localities.

1. The Dalles (at 5 mile rapids), Wasco Co. Recent. No Fm.
   Refs. 138d.
2. Fossil Lake, Lake Co. No Fm. Pleistocene. Refs. 4,
   41, 41m, 85, 85a, 92, 138, 138a, 188, 188a, 188b, 234.
3. Juntura, Malheur Co. Juntura Fm. Miocene (Clarendonian).
   Refs. 25a.
4. McKay Reservoir, Umatilla Co. ?Deschutes Fm. Miocene
   (Hemphillian). Refs. 25.
5. Paulina Creek valley, Crook Co. ?Mascall Fm. Miocene
   (Barstovian). Refs. 188c, 234.
6. Rome, Malheur Co. (5 mi. SW of Rome on Dry Creek). No Fm.
   Miocene (Hemphillian). Refs. 25a, 138c.
7. Sunset Bay, Coos Co. Coaledo Fm. Eocene. Refs. 138b, 234.
8. Willow Creek valley, Malheur Co. No Fm. ?Miocene
   (Hemphillian). Refs. 25a, 188c, 234.

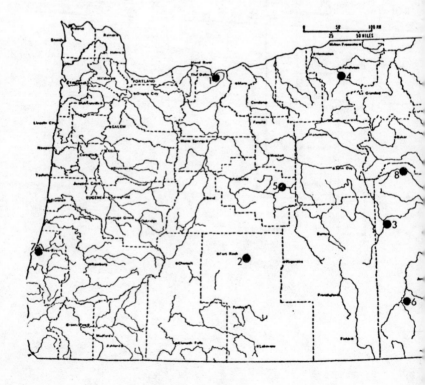

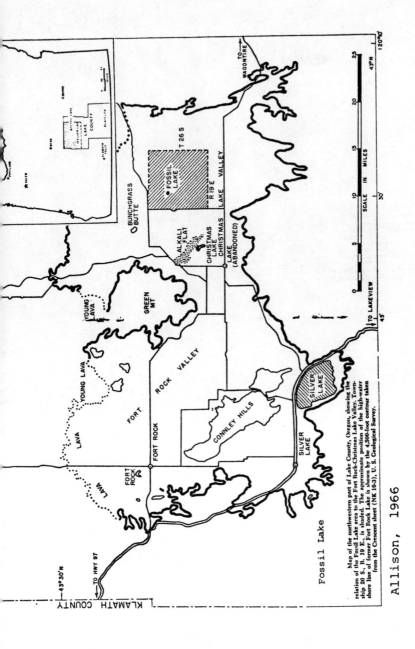

Map of the northwestern part of Lake County, Oregon, showing the relation of the Fossil Lake area to the Fort Rock-Christmas Lake Valley. Township 26 S., R. 19 E., is shaded. The approximate position of the high-water shore line of former Fort Rock Lake is shown by the 4,300-foot contour taken from the Crescent sheet (NK 10-3), U. S. Geological Survey.

Fossil Lake

Allison, 1966

183

OREGON FOSSIL MARINE VERTEBRATES

Grouping the various classes of mammals, sharks, bony fish and reptiles together in this chapter may seem peculiar except from the geologists' point of view. At death and burial, geologic processes combine the remains of organisms from various habitats throughout the ocean environment. Open ocean forms like whales and porpoises are often associated in fossil assemblages with coastal forms such as seals, sea otters, and shore birds. Sharks operating as scavengers are invariably drawn to carcasses prior to geologic burial, and it is not uncommon to find shark teeth still imbedded in the fossilized remains of larger animals.

In spite of the comparative large body size of many of the marine vertebrates, the fossil record is remarkably poor compared to terrestrial forms. This limited fossil record is reflected not only by badly scattered remains but by physical wear or abrasion on the individual fossil bones and teeth as well.

Three primary factors contribute to the dismal fossil record of marine vertebrates. In living populations one very important factor is the limited standing crop. Except for small fish, most marine vertebrates are at or near the top of a food pyramid. This means that they will have very small standing crops or populations. Even if an animal is physically large, numerical advantage is what really counts with respect to successfully getting into the fossil record. If a large whale does become preserved the probability of that single specimen being found by a geologist or paleontologist is remote. The efficiency of scavengers in the marine environment cannot be overlooked as a detriment to fossil preservation. An impressive array of carrion feeders from large sharks to crabs and small fish await the opportunity of any carcass large or small. One of the effects of scavengers is to thoroughly scatter the bones of a skeleton. Finally the marine environment itself may be the most important detriment to vertebrate fossil preservation. Large carcasses washing ashore as well as the remains of coastal animals carried into the ocean are subject to intensive sorting, dispersion and abrasion by the high energy of the surf zone. Although we normally think of the ocean as an environment of deposition, the beach wear on bone material may often preclude an effective burial for an animal's appearance in the fossil record. Another factor worth noting is the virtual lack of specialized preservation/trapping environments in the ocean. This is in reference, of course, to situations familiar to us on land such as quicksand, tar pits, and bogs that trap terrestrial vertebrates and preserve them intact often in an articulated condition. In some cases, particularly with peat bogs, the chemistry of the stagnant water may contribute to preserving even hide, hair and flesh. On a smaller scale, but no less important, pitch on trees traps and preserves all manner of insects, spiders, and pollen.

CETACEA

Fossil and living members of the Order Cetacea, or whales and porpoises, are divided into three major groups or suborders:
1. *Archaeoceti* or extinct primitive toothed whales.
2. *Odontoceti* or modern toothed whales and porpoises.
3. *Mysticeti* or baleen "whalebone" whales.
The two former groups are carnivorous predators occupying the top positon in food pyramids. The latter, by contrast, have essentially eliminated the middle man in a food pyramid by feeding directly on plankton. Streamlining in body form reflects the individual species' lifestyle and may be used in evaluating fossil specimens to extrapolate the same data. Almost all modern large whales are baleen types. Despite the fact that Cetacea are only poorly known from the Oregon fossil record, all three of the above mentioned suborders are represented.

The first whales or cetaceans appeared over 50 million years in the past during the Eocene epoch. These Eocene Archaeocetes or primitive toothed whales disappear by the end of the Oligocene time, but not before they give rise to the Odontocetes and Mysticetes during the same period.

The first fossil cetacean to be recorded in Oregon was by J. Dana with the Wilkes Expedition of 1838-1842 when he recovered bone fragments and vertebra from the Astoria Formation near Astoria about 13 miles above the mouth of the Columbia River. The fragments were found in association with molluscan and arthropod remains. Compared to California, the Oregon cetacean fossil record has received very little attention in the scientific literature. This may be remedied when the large collection made by Douglas Emlong in the late 1950's and presently housed in the Smithsonian Institution in Washington, D. C., is prepared and described. Of more than 1,000 separate field numbers assigned to this collection, almost half apply to cetacean remains. The remainder of the Emlong collection is made up of seals, sea lions, and vertebrates in other Orders.

Several Oligocene formations in the State have yielded cetacean remains. Best known of these is the Yaquina Formation near Newport, Oregon. Emlong's (1966) description of a primitive whale, *Aetiocetus cotylalveus*, from that site included the complete skull and most of the vertebra. This animal belongs to the Suborder Archaeoceti and was around twelve feet in length. Further inland near Silverton, Oligocene shoreline deposits of the "Butte Creek Beds" bear sparce remains of apparently the same species as from the Yaquina Formation.

At least two other varieties of Archaeoceti skulls have been collected by Emlong from the late Oligocene Alsea Formation. The paleoenvironmental setting of several other Oligocene Formations in the State including the Eugene Formation, the Pittsburgh Bluff Formation and the Keasey Formation suggest that these units should yield cetacean remains, but to date none have been reported.

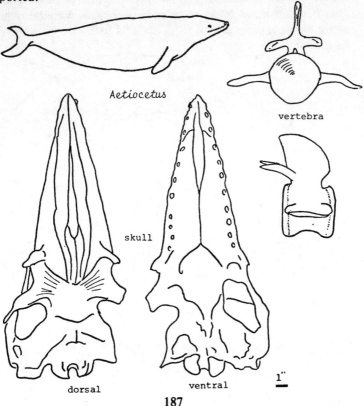

Aetiocetus

vertebra

skull

dorsal   ventral   1"

187

Many individual fossil specimens and fragments of Cetacea have been recovered from localities along the coast. Two of these are notable: 1. The above mentioned middle Miocene Astoria Formation has long been known for its rich molluscan fossils and commonly yields cetacean remains usually in the form of vertebra. Commonest is the primitive cetothere, *Cophocetus oregonensis* (baleen whale) described by Packard and Kellogg (1943). Packard (1940) mentions an undescribed porpoise from the Oregon Miocene but gives no other specific information. 2. The late Miocene Empire Formation, particularly in the vicinity of Fossil Point near Coos Bay, is a source of cetacean bone material primarily represented by ribs or vertebra. Although many specimens have been collected from both Formations, the last paper on Oregon Pliocene or Miocene Cetacea appeared over thirty years ago.

In addition to cetaceans in the Miocene Empire and Astoria Formations, Pliocene localities near Cape Blanco have yielded cetacean material. A remarkable find in this regard was noted by Packard (1947) where he described fossil baleen from a Mysticete whale. Baleen is chitonous in composition and is unlike bone in that it is much less likely to be preserved. This material requires rapid burial and protection from bacteria to ensure its occurrence as a fossil.

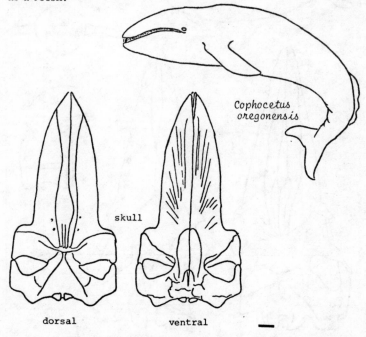

Cophocetus oregonensis

skull

dorsal   ventral

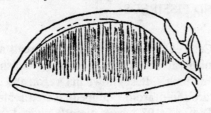

Skull of Mysticete whale showing baleen

It is curious that although two new groups, *Mysticeti* and *Odontoceti*, appear for the first time in the Oligocene epoch, this interval in time is notable for its particularly low diversity of cetaceans. Several authors have considered this mid-Tertiary lull in cetacean diversity and have been able to show a similar diversity lull in the plankton over the same time interval. Although the baleen whales (*Mysticeti*) are more directly tied to plankton, there can be little doubt of the key role of plankton in the food pyramid of the toothed whale. Further evidence suggests that this lull in plankton populations and diversity may have been triggered by slow but profound changes in ocean temperatures and upwelling patterns throughout the same time interval. Elsewhere in the fossil record it is not uncommon for a "crisis environment" to trigger the development and appearance of several new and innovative life forms.

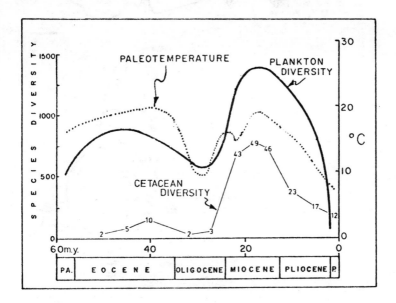

PINNIPEDS AND FISSIPEDS

Unlike the Cetacea, the fossil pinnipeds or seals and sea lions have received considerable attention in recently published literature on fossil marine mammals. Best known in Oregon and elsewhere on the West Coast are the otaroid pinnipeds (eared seals; sea lions). Repenning and Tedford (1977) have recently revised the taxonomy of this group and note two Oregon species, *Pontolis magnus* (Dall, 1909), from the Miocene Empire Formation at Coos Bay and *Desmatophoca oregonensis* from the Miocene Astoria Formation at Newport (Packard, 1947). In addition to complete skulls, specimens of limb bones and vertebra are reported elsewhere from the same Formations. Barnes and Mitchell (1975) report skull fragments of a small young adult female seal, *Phoca* cf. *P. vitulina* Linnaeus, from late Tertiary rocks in the sea cliffs at Cape Blanco from either the Empire or Port Orford Formations. Packard (1947)

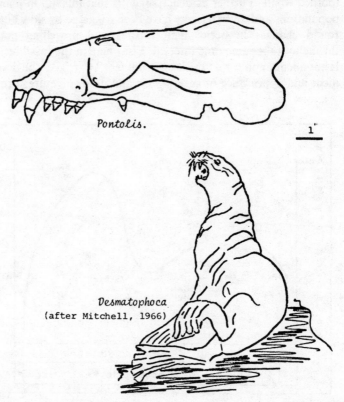

Pontolis.

1"

Desmatophoca
(after Mitchell, 1966)

190

reports the only sea lion to be found to date of the genus *Eumetopias* from Oregon Tertiary rocks. The specimen was collected by E. M. Baldwin from the mouth of the Elk River south of Cape Blanco, and bone material consists of a single radius. Regarded as Pliocene by Packard, this specimen is similar to the living northern sea lion, *Eumetopias*, still present today off the Oregon coast.

Leffler (1964) reviews coastal vertebrate finds from the upper Elk River Formation, Curry County. Here cheek teeth of a tapir reported by Merriam (1913) and an undescribed otarid (eared) seal collected by B. Martin in 1911 occur near the same level as a sea otter (fissiped) femur. The sea otter find, *Enhydra* sp., in the late Pliocene/early Pleistocene represents the earliest known occurrence of this genus in North America. In association with these vertebrate finds, the microfossils and megafossils (Zullo, 1961) from this site suggest shallow, offshore, sandy, marine shelf environment in cold water of probably less than 50 feet depth.

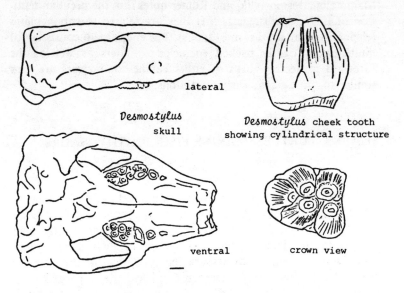

Desmostylus skull — lateral

Desmostylus cheek tooth showing cylindrical structure

ventral

crown view

In addition to pinnipeds and fissipeds, the Oregon Neogene bears fossils of what may be an extinct sirenian or sea cow, *Desmostylus*. A *Desmostylus* tooth found by T. Condon from the Yaquina Bay area was reported by H. Osborn (1902), but it wasn't until 1914 that a second *Desmostylus* discovery was announced and illustrated by E. Condon McCornack. This *Desmostylus*, a skull,

Desmostylus reconstruction (after Mitchell, 1966)

from the Astoria Formation at Spencer Creek was later redescribed by Hay (1916) and again by Vanderhof (1937). Repenning (1977) has referred to this as a four legged sirenian living in the Pacific area from 10-25 million years ago, but some conflict as to its ancestry remains. *Desmostylus* is placed in a separate order of Mammalia, Desmostylia, and Romer notes that the peculiar dentition of this group suggests that they are related to subungulates (elephants, mammoths, mastodons). The cheek teeth consisting of multicusped, tightly packed enameled cylinders as well as the foreward projecting tusks of either canines or incisors are very similar to those of the early mastodons.

SHARKS, TURTLES, MARINE FISH, ICHTHYOSAURS

Three Classes are considered in this section: first is the Class Chondrichthyes, or sharks and rays; second is the Class Reptilia, or turtles and ichthyosaurs; third is the Class Osteichthyes, or bony fish. Shark remains in Oregon are known from Cretaceous, Eocene, Oligocene, and Miocene. Pliocene and Pleistocene Formations as yet have no known shark remains. Most of these occurrences are of fossil teeth. Dermal denticles, the tiny parts of a shark's armored skin, often turn up in processed microfossil samples, but to date none have been described. Because their cartilaginous skeleton deteriorates too rapidly for preservation and because of continual tooth replacement throughout their lifecycle, almost all shark remains in Oregon are teeth. Tooth enamel is sufficiently hard and chemically inert to resist the wear, abrasion and solution of geologic processes. Oregon fossil sharks are primarily groups that had their origin in the Mesozoic. The earliest shark remains in Oregon from the Cretaceous Hudspeth Formation near Mitchell are noted by Welton (1972) in his review of Oregon fossil sharks. One of the earliest works on fossil sharks is Jordan's (1923) description of three Eocene species collected from the Coaledo Formation. In one Eocene locality at Scoggins Creek in the Yamhill Formation vertebra and even calcified cartilage have been reported by Applegate (1965). In a thorough study of mid Oligocene molluscs from the Eugene Formation Hickman (1969) reported both galeoid and the saw-like teeth of the primitive hexanchoid sharks.

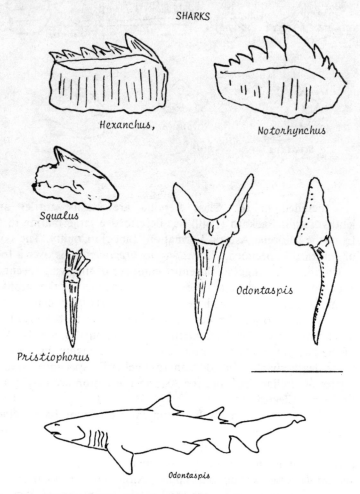

SHARKS

Hexanchus,

Notorhynchus

Squalus

Odontaspis

Pristiophorus

Odontaspis

193

Rays and skates are bottom dwelling fish feeding primarily on invertebrates. Both groups are characterized by a pavement-like dental surface in the mouth adapted for crushing mollusk shells. Exposures of the Coaledo Formation at Shore Acres State Park near Charleston commonly yield fragments and complete teeth of the Eagle Ray, *Myliobatis*. According to Welton, this is the commonest shark tooth type in Oregon.

SHARKS, SKATES AND RAYS

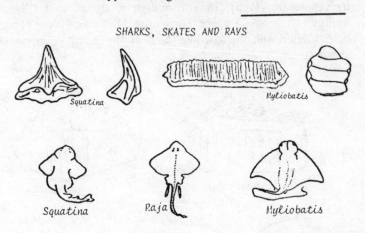

Included in the Class Reptilia are marine turtles and ichthyosaurs. Packard (1940) has described a large marine turtle from the Miocene Astoria Formation, Lincoln County. The skull of Packard's specimen, *Psephophorus oregonensis*, is over a foot in length, which suggests the entire carapace of the living specimen may have been well over six feet long. A fragment of the carapace from the same locality includes some twenty-six bony plates fitted together in a mosaic. The family to which this turtle belongs, Dermochelydae, includes the modern leatherback turtles, but the skull of the Oregon specimen is more narrow and elongate than that of living leatherbacks. In addition to Packard's specimen, several turtles of similar size from the Astoria Formation are part of the Emlong Collection.

Mesozoic units in southern and eastern Oregon have yielded occasional specimens of the reptile ichthyosaur (Merrian and Gilmore, 1928; Camp and Koch, 1966). These large and distinctive marine reptiles are several orders older than any of the aquatic vertebrates discussed up to this point. Although their occurrence in Oregon is neither prolific nor particularly well-preserved, they are

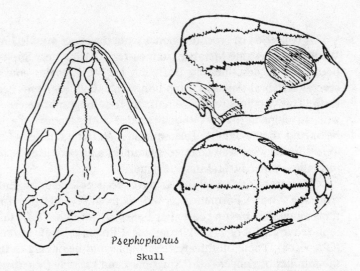

Psephophorus
Skull

mentioned here to illustrate the considerable breadth of the Oregon fossil record. The sleek streamlined appearance of these marine reptiles attests to their complete adaptation to the aquatic environment. Rows of sharp conical teeth on a beaked jaw suggest that the ichthyosaur was an aggressive predator on fish. Cervical vertebral elements from mid Cretaceous exposures in Wheeler County as well as rostrum and tooth elements from Jurassic exposures in Curry County are from animals around fifteen feet in length. Elsewhere complete specimens from Triassic exposures in the western U. S. are as much as sixty feet in length. Marsh (1895) has a brief reference to an ichthyosaur vertebra found in the Blue Mountains in association with the Mesozoic pelecypod genus, *Trigonia*, but gives no further data.

anterior view of cervical
(neck) vertebra

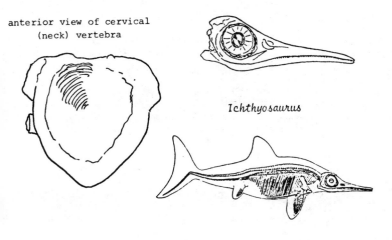

Ichthyosaurus

195

A collection of crocodile bones from the lower middle Jurassic Wedberg Limestone Member from eastern Oregon near Suplee has been recently described by Buffetaut (1979). Remains consist of several vertebral elements, limb bones and skull fragments bearing teeth. This was apparently a small, six foot long, marine crocodile with an elongate narrow snout studded with slender long teeth, belonging to the family Teleosauridae. Buffetaut notes that the Teleosaurids were well adapted swimmers and were usually found in shallow water littoral marine deposits.

Marine fish, Osteichthyes, have been collected by Emlong from the Astoria Formation as well as from several other formations along the Oregon coast. Ear bones or "otoliths" of fish were reported by Moore (1976) from the Pittsburg Bluff Formation (Oligocene). These otoliths were recognized as belonging to fish in the families of conger eels (Congridae) and rat tails (Macruridae). Lore Rose David (1956) reports fossil scales of anacanthin fish from Eocene and Oligocene units of western Oregon including the Yaquina, Keasey, Nestucca and Nye Formations. Anancanthin fish

ANACANTHIN MARINE FISH SCALES (after L.R. David, 1956)

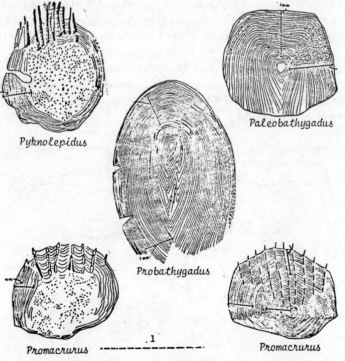

Pyknolepidus

Paleobathygadus

Probathygadus

Promacrurus

Promacrurus

including such modern forms as cod and haddock are referred to by David as the "herrings of the deep sea" because of their abundance in deep water. Although David's work considers only fish scales, the fingerprint like pattern of grooves and ridges on the face of each scale is consistant and reliable as a character for positive identification of species. Of particular significance to Oregon paleogeography is David's interpretation of a "bathyal" paleoenvironment for each of the above five formations at the levels at which the scales were found. This interpretation agrees in broad form to paleodepth estimates previously made on the same formations on the basis of foraminiferal and molluscan fossil assemblages.

OREGON FOSSIL MARINE VERTEBRATE CLASSIFICATION

The following classification contains Oregon fossil
marine vertebrate genera listed in the published
literature. Taxonomic revision has not been attempted.

CLASS: Mammalia
  ORDER: Cetacea
    SUBORDER: Archaeoceti
      FAMILY: Aetiocetidae (primitive whale)
        GENERA: *Aetiocetus.*
    SUBORDER: Odontoceti
    SUBORDER: Mysticeti
      FAMILY: Cetotheriidae (baleen whale)
        GENERA: *Cophocetus.*
      FAMILY: Balaenidae
  ORDER: Desmostylia
      FAMILY: Desmostylidae
        GENERA: *Desmostylus*
  ORDER: Carnivora
    SUBORDER: Fissipedia
      FAMILY: Mustelidae (sea otter)
        GENERA: *Enhydra*
    SUBORDER: Pinnipedia
      FAMILY: Phocidae (earless seal)
        GENERA: *Phoca.*
      FAMILY: Otaridae (eared seal, sea lion)
        GENERA: *Desmatophoca, Eumetopias, Otarid,*
                Pontolis.
CLASS: Reptilia
  ORDER: Ichthyosauria
      FAMILY: Ichthyosauridae
        GENERA: *Ichthyosaurus*
      FAMILY: Dermochelydae (turtle)
        GENERA: *Psephophorus*
CLASS: Osteichthyes
  ORDER: Anguilliformes
      FAMILY: Congridae (eel)
  ORDER: Gadiformes (Anacanthini)
      FAMILY: Bathygadidae (deep-water fish)
        GENERA: *Paleobathygadus, Probathygadus*
      FAMILY: Macruridae (deep-water fish)
        GENERA: *Calilepidus, Promacrurus, Pyknolepidus*
CLASS: Chondrichthyes
    SUBORDER: Batoidea (rays and skates)
      FAMILY: Rhinopteridae
        GENERA: *Rhinoptera*
      FAMILY: Myliobatidae
        GENERA: *Myliobatis, Aetobatus*
      FAMILY: Rajidae
        GENERA: *Raja*

```
CLASS: Chondrichthyes
    SUBORDER: Selachii (true sharks)
        FAMILY: Carchariidae
            GENERA: Carcharias, Odontaspis, Scapanorhynchus
        FAMILY: Isuridae
            GENERA: Carcharodon, Isurus, Lamna
        FAMILY: Squalidae
            GENERA: Centroscymnus, Centrophorus, Squalus
        FAMILY: Squatinidae (Rhinidae)
            GENERA: Squatina
        FAMILY: Heterodontidae
            GENERA Heterodontus
        FAMILY: Hexanchidae
            GENERA: Heptranchias, Hexanchus, Notorhynchus
        FAMILY: Pristiophoridae
            GENERA: Pristiophorus
        FAMILY: Carcharhinidae
            GENERA: Galeocerdo
        FAMILY: Sphyrnidae
            GENERA: Sphyrna
        FAMILY: Echinorhinidae
            GENERA: Echinorhinus
        Family: Scyliorhinidae
            GENERA: Scyliorhinus
```

199

OREGON MARINE VERTEBRATE BEARING FORMATIONS

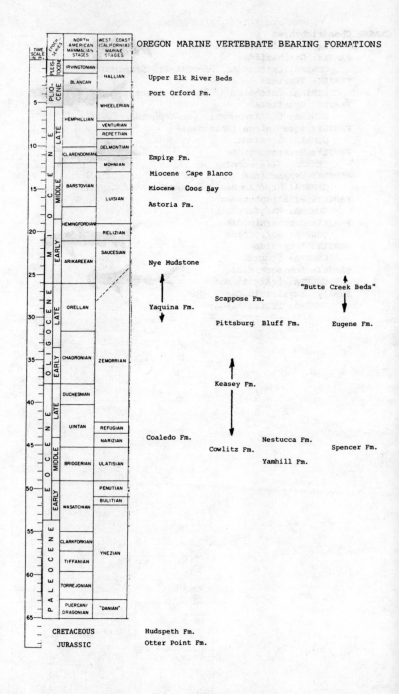

OREGON FOSSIL MARINE VERTEBRATE RANGES

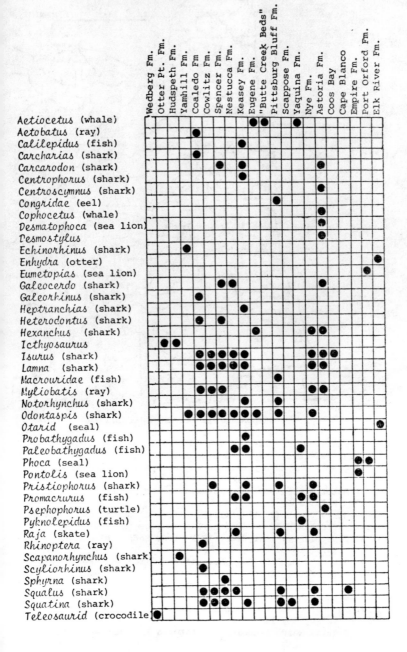

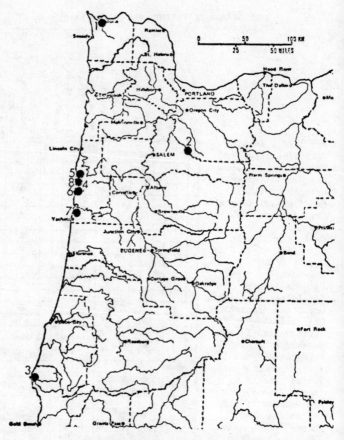

OREGON FOSSIL CETACEAN LOCALITIES

Following is a list of Oregon fossil cetacean localities.
The reference numbers are to authors listed in the bibliography who have published on the specific localities.

1. Astoria, Clatsop Co. Astoria Fm. Miocene.
   Refs. 236.
2. Butte Creek (Scotts Mills) valley, Clackamas Co. "Butte
   Creek Beds". Oligocene/Miocene. Refs. 152.
3. Cape Blanco, Curry Co. ?Elk River Beds. Pliocene.
   Refs. 155b.
4. Newport, Lincoln Co. Astoria Fm. Miocene. Refs. 158.
5. Otter Rock, Lincoln Co. Astoria Fm. Miocene.
   Refs. 158.
6. Seal Rock, Lincoln Co. Yaquina Fm. Oligocene. Refs. 55.
7. Spencer Creek, Lincoln Co. Astoria Fm. Miocene.
   Refs. 78b, 142, 158.
8. Yaquina Head, Lincoln Co. Astoria Fm. Miocene. Refs. 158.

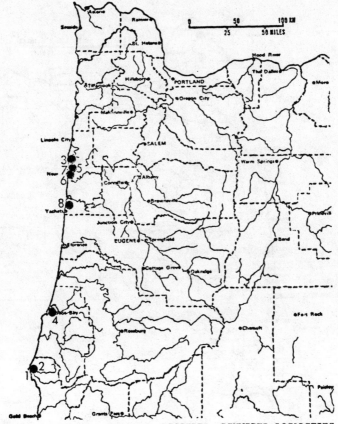

OREGON FOSSIL DESMOSTYLUS, FISSIPED, PINNIPED LOCALITIES

Following is a list of Oregon fossil desmostylus, fissiped, and pinniped localities. The reference numbers are to authors listed in the bibliography who have published on the specific localities.

1. Cape Blanco, Curry Co. Elk River Beds. Pliocene/Pleistocene.
   Refs. 109.
2. Cape Blanco, Curry Co. Port Orford Fm. Pliocene.
   Refs. 14, 155d.
3. Cape Foulweather, Lincoln Co. ?Astoria Fm. Miocene.
   Refs. 73.
4. Coos Bay, Coos Co. Empire Fm. Miocene. Refs. 43, 186.
5. Johnson Creek, Lincoln Co. Astoria Fm. Miocene.
   Refs. 155e.
6. Newport, Lincoln Co. Astoria Fm. Miocene. Refs. 38.
7. Schooner Creek, Lincoln Co. Astoria Fm. Miocene.
   Refs. 155e.
8. Spencer Creek, Lincoln Co. Astoria Fm. Miocene. Refs. 78b.

OREGON FOSSIL SHARK LOCALITIES

Following is a list of Oregon fossil shark localities. The
reference numbers are to authors listed in the bibliography
who have published on the specific localities.

1. Astoria, Clatsop Co. Astoria Fm. Miocene. Refs. 236.
2. Cape Blanco, Curry Co. No Fm. Miocene. Refs. 233.
3. Coos Bay, Coos Co. Coaledo Fm. Eocene. Refs. 97, 233.
4. Eugene, Lane Co. Eugene Fm. Oligocene. Refs. 80, 201c.
5. Helmick Hill, Polk Co. Spencer Fm. Eocene. Refs. 181, 233.
6. Mist, Columbia Co. Keasey Fm. Eocene/Oligocene. Refs. 233.
7. Mitchell, Wheeler Co. Hudspeth Fm. Cretaceous. Refs. 233.
8. Nehalem River valley, Columbia Co. Cowlitz Fm./Pittsburg
    Bluff Fm. Eocene/Oligocene. Refs. 201b, 233.
9. Newport, Lincoln Co. Astoria Fm. Miocene. Refs. 233.
10. Scoggins Creek valley, Washington Co. Yamhill Fm. Eocene.
    Refs. 6.
11. Toledo, Lincoln Co. Nestucca Fm. Eocene. Refs. 233.

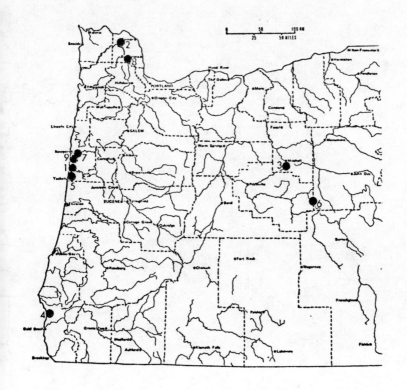

OREGON TURTLE, ICHTHYOSAUR, ANACANTHIN, AND CROCODILE LOCALITIES

Following is a list of Oregon turtle, ichthyosaur, anacanthin
and crocodile localities. The reference numbers are to authors
listed in the bibliography who have published on specific
localities.

1. Alsea Bay, Lincoln Co. Yaquina Fm. Oligocene. Refs. 45.
2. Mist, Columbia Co. Keasey Fm. Eocene. Refs. 45.
3. Mitchell quadrangle, Wheeler Co. No Fm. ?Cretaceous.
      Refs. 136.
4. Sisters Rocks, Curry Co. ?Otter Point Fm. Jurassic.
      Refs. 105.
5. Spencer Creek, Lincoln Co. Astoria Fm. Miocene. Refs. 155a.
6. Suplee, Crook and Grant Co. Wedberg Member (Snowshoe Fm.).
      Jurassic (Bajocian). Refs. 28.
7. Toledo, Lincoln Co. Nestucca Fm. Eocene. Refs. 45.
8. Vernonia-Scappose Road, Columbia Co. Pittsburg Bluff Fm.
      Oligocene. Refs. 142b.
9. Yaquina Bay, Lincoln Co. Nye Mudstone. Miocene. Refs. 45.

205

OREGON FOSSIL MAMMALS
AND OTHER LAND VERTEBRATES

Mammals first appear in the fossil record during the Triassic period over 200 million years in the past. Early mammals were small rat-size organisms that co-existed with dinosaurs for several million years until the latter group gradually disappeared near the end of the Cretaceous period. Although Triassic rocks do occur in Oregon, most of these are exposures of marine rocks not likely to bear mammalian fossils. Similarly, rocks of the following periods, the Jurassic and Cretaceous, occur in Oregon, but again these are primarily marine in origin. The oldest mammal-bearing fossiliferous rocks in Oregon are Eocene in age.

When the geologic record was being delineated and standardized in the early 1800's the research was carried out in Britain and western Europe. Europe during the Tertiary period was a series of small marine inlets or basins off the sea. Time terms like "Miocene," "Eocene," and "Pliocene" were defined in distinctive exposures of sediments in Europe bearing characteristic marine molluscan fossils. A complete chronology was ultimately established, and these European time units could even be recognized in the major marine sediment basins in North America. Nonmarine rocks bearing mammalian fossils were at that time impossible to correlate accurately due to the totally different environments involved.

Fossil mammals occur in nonmarine rocks including wind blown dune sands, fine-grained lake deposits as well as in the coarse sand and gravel deposits of streams and rivers. Although occasional mammals other than cetaceans (whales) or pinnipeds (seals) are found in marine rocks, we do not normally regard marine units as probable areas for extensive mammalian vertebrate fossil remains. In some areas, marine rocks can be traced laterally directly into nonmarine units of the same age. In such cases we have traversed a prehistoric "strand" or shoreline between the old ocean and land areas. Where marine and nonmarine units interfinger in this manner we can correlate or "age date" the nonmarine animals by the molluscs in the adjacent marine units. For the most part, however, the terms "Miocene," "Eocene," etc., held little meaning for the paleontologist bent on correlating or dating nonmarine units with mammalian fossils. To remedy this problem, vertebrate paleontologists established their own chronology of "mammalian stages" based on local North American vertebrate faunas. Later, by the use of radioactive decay techniques ("absolute dating"), geologists were able to correlate more precisely these newer mammalian stages with the established European molluscan epochs.

Because of their large populations, the smaller mammals such as gophers, rats, mice and shrews leave a better fossil record than the less numerous ungulates or hoofed mammals. Ungulates in turn invariably leave a better fossil record than the more rare predators. Finally, plains dwelling animals leave a better record than those in a hilly or montane environment because the former is an area of deposition and the latter of erosion. Mammals are represented in the fossil record by bones and teeth which are preserved intact. The cheek or jaw teeth of mammals are highly distinctive in the appearance of the crown surface. The structure of the enamel and dentine along with normal wear on the teeth produces a characteristic pattern for most species of mammals much like a fingerprint of a human. In addition to aiding in the identification of a given mammal, the teeth along with the skeleton reflect the animal's niche in the environment. The latter information is, of course, of fundamental importance in reconstructing prehistoric environments.

MAMMAL CHEEK TEETH (crown view)

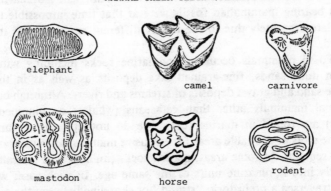

elephant

camel

carnivore

mastodon

horse

rodent

One characteristic of mammals is the rapid evolution of the group. Genera and species of mammals tend to have very short stratigraphic ranges. The duration, then, of mammalian taxa in the geological record from first appearance to disappearance is notably short when compared to fossil plant and invertebrate taxa. These short stratigraphic ranges imply that the mammals evolve more rapidly than other groups but, more important, they dictate that once a mammal fossil is identified, the age of the entombing rock is pinpointed particularly well. The average invertebrate organism often leaves a single fossil specimen. A clam, for example, may be separated into the two separate valves in the fossil record. Plants leave tens of thousands of fossil specimens, but here a problem arises in associating a particular species of pollen grain with a species of leaf once they have become separated and scattered in the fossil record. A single vertebrate animal may leave over one hundred separate fossil bones, but the complexity of the mammalian skeleton is such that the various parts of the skeleton can be readily associated with a given species.

The condition and frequency of the mammalian fossil skeletal elements may provide insight on the environmental conditions at the site of burial and entombment. Shotwell (1958) has noted that if a representative number of bones of a species are found at one site, original burial probably occurred at that site and the bone material did not suffer transport by geologic agents such as running water. Specimens found represented by only one or a very few bones at a given site might be regarded as having been transported into that environment after death by either running water or possibly by a predator/scavenger.

CLARNO FORMATION MAMMALS

Eocene fossil mammal localities are rare in Oregon. The Clarno Formation in the John Day Valley has yielded vertebrate remains, but most of these are freshwater fish. Mammalian remains from the Clarno Eocene include skeletal material and teeth, as well as a complete skull of the genus *Amynodon* (Hancock, 1962), a large aquatic rhino. An upland running rhinoceros, *Hyrachus*, found near the Clarno Bridge was dated as middle Eocene by Stirton (1944). Slender and strikingly pony-like in appearance, this agile looking rhinoceros was about the size of a Great Dane. Large primitive ungulates, including brontotheres, pantadonts and uintatheres as well as tapirs and the chalicothere *Moropus* also occur in the Clarno (Hancock, 1962). *Moropus* was a large browsing chalicothere perissodactyl related to horses but bearing claws instead of hoofs. One bizarre adaptation occuring in the Astoria (Miocene) Formation is a domed skulled chalicothere. Munthe and

Coombs (1979) in describing several of these odd-shaped skulls go to some lengths in considering the adaptive significance of the domes including as an aquatic snorkel, an attachment for jaw muscles, for water retention (like a camel's hump), for filtering or humidifying air, as an increased brain capacity in one area of smell,

210

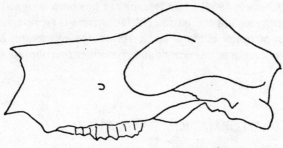

Chalicothere.

(after Munthe and Coombs, 1979)

hearing, etc. They conclude the probable function of the dome is either a butting surface during mating or sexual display or a visual or acoustical signaling device. Horses, crocodiles and oreodonts from the Clarno are mentioned by Shotwell.

A partial skull of the carnivore *Hemipsalodon grandis* was reported from the Clarno Mammal Quarry by Mellett (1969). This skull is nearly 16 inches in length and represents a lineage of very large early Tertiary hyaenodontids of the creodont suborder Deltatherida. Several of the vertebrates from the Clarno Mammal Quarry locality are regarded elsewhere as restricted to Oligocene rocks and the locality is dated as early Oligocene (Chadronian) age. When one considers the richness of the plant materials in parts of the Clarno, it is remarkable that only a few vertebrates are found in the formation.

Baldwin reports absolute (radiometric) dates of 43.1 m.y.b.p. for the lower Clarno Formation and 34.0 m.y.b.p. for the Clarno Mammal Quarry of the upper Clarno. These dates would suggest that the lower Clarno is Duchesnean (late Eocene) grading into the Chadronian (lower Oligocene) in the upper Clarno. The Clarno Mammal Quarry is the only early Oligocene vertebrate faunal locality reported in Oregon. Elsewhere in discussions on fossil

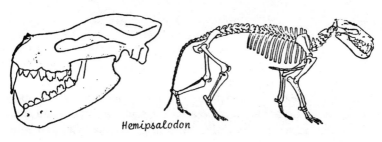

Hemipsalodon

211

plants (Chaney, 1956) a late Eocene age has been assigned to plant floras from the upper Clarno. The latter flora is representative of a moist, subtropical climate, and certainly the vertebrates including varieties of aquatic rhinoceros and crocodiles corroborate this finding.

JOHN DAY FORMATION MAMMALS

Of all the rock units in the State of Oregon, none approach the John Day Formation for richness of fossil assemblages. With exposures located in north central Oregon, the John Day ranges from 700 to 1500 feet in thickness. Three divisions originally described by Merriam (1901) include the lower reddish beds, the middle greenish beds, and the uppermost white beds. Although the lower John Day of early Oligocene (Chadronian) age is synchronous in part with the Clarno Formation (36.0 m.y.b.p.), the two units have not been found interfingering throughout their extent. The upper John Day extends well into the Hemingfordian stage (early Miocene) (18.0 m.y.b.p.), and biostratigraphic zonal division has been developed by Fisher and Rensberger (1972) to correlate the unit with similar rocks in South Dakota. These "zones" are based on distinctive suites of rodents including pocket gophers as well as beaver and a bizarre horned gopher *Mylagaulodon*.

cheek tooth (crown view

Mylagaulodon

Whereas the lower (reddish) John Day is for the most part barren of vertebrate fossils, the middle John Day (greenish) and upper John Day (white) members are rich in well-preserved fossil material. The color distinction of members has now been dropped as time stratigraphic units because the members are found interfingering at several localities. Although Thomas Condon interpreted the John Day to be of lacustrine or lake deposit origin, it is presently regarded to be largely aeolian or wind deposited. While the unit is bounded by unconformities, if hiatuses exist in the John Day they are not apparent as the unit is evidently a continuous uninterrupted record from Chadronian to Hemingfordian. Four members of the John Day Formation are currently recognized (Fisher and Rensberger, 1972). Beginning at the bottom, the Big Basin Member of red claystone bears Oligocene plant remains but very little in the way of vertebrates. Overlying the Big Basin, the Turtle Cove Member is dominantly greenish tuffs bearing an extremely rich vertebrate fauna from Chadronian to Arikareean age. Overlying the Turtle Cove are the buff to gray tuffs of the Kimberly Member bearing rich Arikareean stage mammalian faunas. The uppermost member in the sequence is the mixed aeolian/lacustrine tuffs and conglomerates of the Haystack Valley Member bearing vertebrate faunas correlative with the Hemingfordian age.

Original fossil collections in the middle and upper John Day Formation of the John Day basin were made by Thomas Condon in the late 19th century and later described by several authors. Condon himself developed a geologic history of the area after making extensive collections there as well as studying fossils brought to him by soldiers passing through on their way to The Dalles. Throughout the last century many paleontologists have been attracted to Oregon by the richness of the John Day fossil beds. Summarizing a few of the more important papers provides the reader with some insight into the diversity of the assemblages. The earliest and most prolific author, E. D. Cope, reported on all manner of fossil vertebrates collected in the John Day Formation. He published close to 40 papers including monographs on cats (1880), dogs (1882,1883a), peccary (1880), rodents (1883b), camels (1886) as well as several general papers. Merriam and Sinclair (1907) reviewed earlier papers on the John Day Formation and compiled a composite faunal list. Thorpe (1921-1925) produced a series of well-illustrated monographs on a variety of John Day mammals including artiodactyls and carnivores.

Cats in the John Day Oligocene include the sabre-toothed *Hoplophoneus* and *Dinictis*. These were sleek lightweight predators somewhat larger than an lynx. The primitive dog *Cynodictis* is also known from this unit. About the size of a fox, *Cynodictis* was almost weasel-like in the proportions of its long tail and body. Oreodonts are well represented in the lower John Day Formation and were usually the size of modern sheep. These curious animals

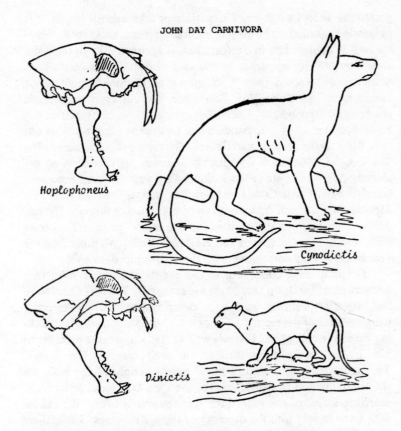

Hoplophoneus

Cynodictis

Dinictis

most often have a wedge-shaped skull similar to those of pigs and peccarys. Often referred to as "ruminant hogs", the oreodonts bear selenodont cheek teeth, whereas pigs and peccaries have bunodont or low crown, blunt cusp molars.

A detailed monograph by Schultz and Falkenbach (1968) on oreodonts of the John Day Formation includes a great deal of attention to the details of stratigraphic distribution as well as taxonomy. Tapirs are not well-known from the Eocene and early Oligocene of North America, and their abrupt appearance in the fossil record here may reflect a migration from Europe. The John Day Oligocene genus *Protapirus* was about half the size of the living South American tapir, *Tapir americanus*. The striking resemblance of *Protapirus* to primitive rhiniceros reflects the closeness of those two perissodactyl lineages.

Squirrels of some 14 species are reported from the John Day

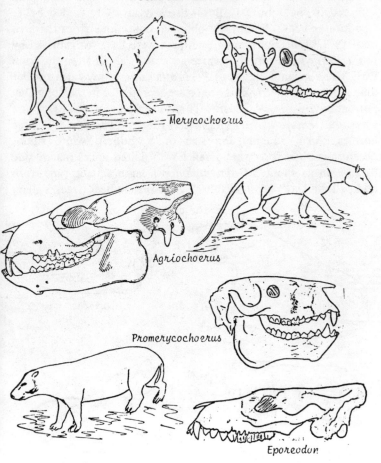

Merycochoerus

Agriochoerus

Promerycochoerus

Eporeodon

Formation and other late Tertiary formations by Black (1963) in localities ranging from Arikareean to Hemphillian in age. Cope (1879) noted the occurrence of rabbits (lagomorphs) in the John Day. In all the Oregon Tertiary fossil record only three genera of rabbits are reported. Romer (1971) notes that they appear well along in the Tertiary although they are enormously successful today.

The Turtle Cove Member is by far the most fossiliferous within the John Day. Several papers deal specifically with assemblages from this member alone. One of the earliest of these after Cope's and Merriam's original work was a monograph by Eaton (1922) on carnivora wherein he reported some three genera

215

of cats from the Marsh Collection at Yale University. Marsh had collected in the John Day Basin as early as 1871, guided by T. Condon, and he subsequently obtained collections from the area until 1877. Collections made by Marsh were later re-examined by Lull (1921) who reported on three genera of camels from the John Day. These include the genus *Poebrotherium*, a tiny form smaller than a domestic sheep. Rhinoceros are represented in the John Day Formation by several genera. Three examples of these are the *Caenopus*, *Diceratherium* and *Metamynodon*. The first is a large hornless form roughly the size of a modern white rhino. *Diceratherium* is somewhat smaller with paired horns placed side by side on its snout. *Metamynodon* is a large aquatic type from middle John Day "green member" exposures in Bear Creek Valley,

JOHN DAY UNGULATES

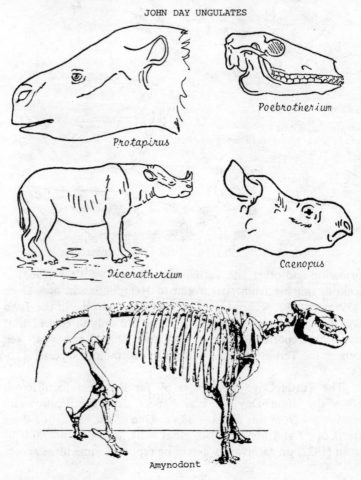

Protapirus

Poebrotherium

Diceratherium

Caenopus

Amynodont

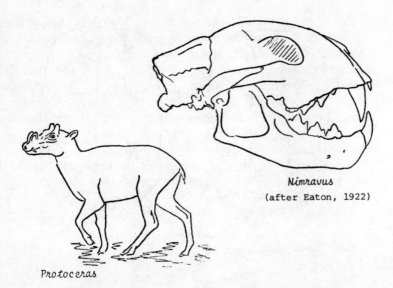

Nimravus
(after Eaton, 1922)

Protoceras

Crook County (Lowry, 1940). Cervids (deer) and the deer-like pecora are also found in the John Day Formation but are not as diverse or abundant as other artiodactyl groups. Several of the cervids are as small as a domestic dog but appear to have been as fleet as they were diminutive. Both pecora and cervoids characteristically bore antlers or bony outgrowths on the skull exemplified by *Protoceras*. In addition, several of the forms in both groups developed dagger-like canines closely resembling the modern day hypertragulids.

Although turtle bones are not uncommon in many Tertiary sites having mammal bones and fossil leaves, only a few of these have been described. Fragments of turtle skeletons have a distinctive shape but are difficult to identify with confidence unless most of the carapace is present. Hay (1903; 1908) reviews two species of turtles from eastern Oregon collections made by Leidy (1871) and Cope (1883e). Although complete turtle carapaces have been recovered from the John Day formation, these reptiles are represented primarily by small fragments of scutes. One exception was the species *Stylemys capax*, Hay, from the John Day which consists of the entire two foot long carapace. *Clemmys hesperia* is described from the Mascall and Rattlesnake Formations in north central Oregon. Two late Miocene giant tortoise carapaces were recovered from material near Arlington in Alkali Canyon (Fry, 1973), and a Miocene turtle, *Clemmys* sp., has been reported from Hemphillian rocks near Rome (Brattstrom and Sturn, 1959).

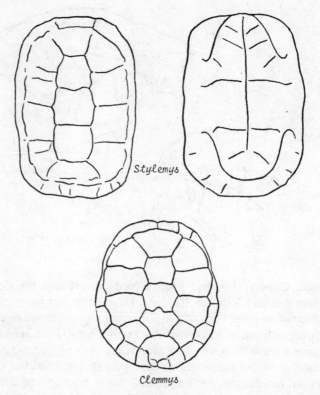

Stylemys

Clemmys

Cope (1883e) reported an Oligocene snake, *Ogmophis*, from the John Day River area. That specimen represented only by vertebra was of the same average size as those living in the area today. Except for turtles most terrestrial reptiles and amphibians have a poor fossil record. The fragile skeleton and the poor environment for preservation do little for the probability of a snake being preserved as a fossil. Berman (1976) has described a new genus and species of burrowing fossorial lizards from the Oligocene/Miocene transition interval within the John Day Formation. The group, Amphisbaenia, is characterized by a wedge-shaped skull with tightly bound bone elements that form the burrowing organ. Living forms of this group are tropical with no limbs and a worm-like body. John Day specimens consist of tiny, one inch long well preserved skulls, one with the lower jaw intact, scattered vertebra and ribs. Fossil material was recovered in a discrete area of the John Day lithology which bears large numbers of fossils of burrowing gophers.

As with amphibians, the fragile nature of frog and toad

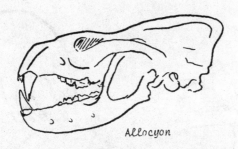

Allocyon

skeletons makes them among the rarest of fossil vertebrates. Nevertheless and Oligocene Fisher Formation locality near the town of Goshen in Lane County periodically yields complete salamander remains in association with fossil leaves. A detailed description and illustrated paper by VanFrank (1955) deals with one such salamander, *Palaeotaricha oligocenica*, Frank. The skeleton is just under 5 inches long and occurs in an articulated condition in a remarkably well-preserved state.

Merriam (1930) described a very dog-like procyonid *Allocyon* from the Crooked River area of Logan Butte, but no detailed stratigraphy was included in the report. Procyonids (racoons and related forms) evolved from the canid (dog) stock in mid-Tertiary time, and Merriam's specimen is a nice transitional type between the two groups. Merriam himself had been studying the John Day area since an early expedition in 1899. His subsequent publications stimulated important research by later students. The transition from dogs to procyonids is also noted by J. C. Wortman (1899) and his John Day material includes six dogs and a mustelid.

In a well-illustrated paper, Sinclair (1905) described several rodents and ungulates, including horses and pigs, from the John Day comprising both Oligocene and Miocene forms. In addition to oreodonts, John Day Oligocene and Miocene rocks have yielded a particularly nice suite of pigs, peccaries, and the giant "pig-like" entelodonts. Some of the members of the latter group were as large as a Shetland pony. On the skull, a well-developed lateral flange below the orbit and pronounced sagittal crest attest to the powerful jaws of the entelodonts.

Haystack Valley Member rodents of early Miocene age are described by Rensberger (1971, 1973). Of special significance is his chronology of faunal replacement of pocket gophers. So often in paleontology it is apparent that one species replaced another, but no good evidence is found to suggest that a species actually dis-

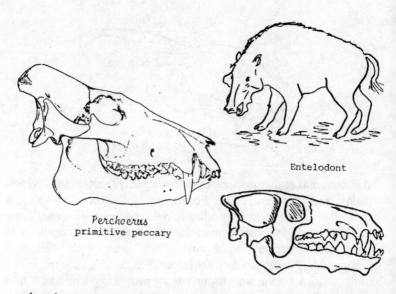

Entelodont

Perchoerus
primitive peccary

placed another through competition. Rensberger's study and his subsequent zonal scheme of gophers support a displacement scheme.

One remarkable anomaly in John Day faunas is the lack of published data on insectivores. This group is particularly important to mammal evolution because the insectivora are the "stem group" (Insectivora) from which other placental mammals were derived. Living shrews, moles and hedgehogs bear many primitive anatomical features that suggest the ancestral position of this group. Insectivores are known from rocks as old as Cretaceous and elsewhere in the world have a persistent though not overly spectacular record through the Tertiary. Stirton and Rensberger (1964) described a new species of the genus *Micropternodus, M. morgani,* from the John Day Miocene. The only reported bat from the John Day Oligocene is a skull and skeletal material described by Brown (1959) in association with plant material.

Woodburne and Robinson (1977) regard the Warm Springs fauna as the youngest material in the John Day Formation. The Warm Springs fauna includes a diverse assemblage of ungulates, carnivores and rodents and is dated as middle Hemingfordian (early Miocene) by those authors. Several of the Warm Springs genera were reported by the latter authors from the John Day Formation for the first time. John Day deposition persisted in the Warm Springs area to the west of the major John Day basin for a short period after the latter had ceased to be a major deposition site.

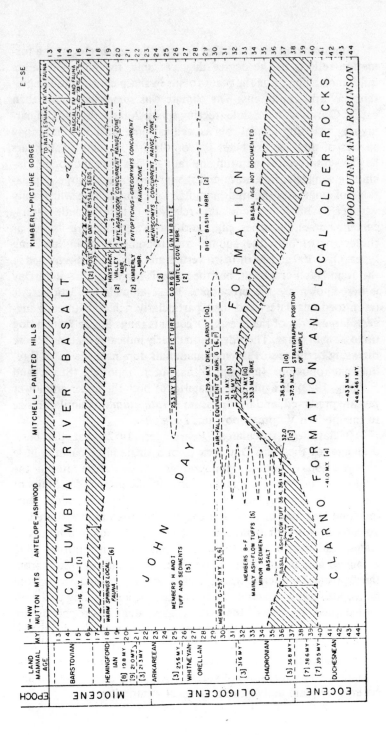

221

Summarizing the John Day fossils, the vertebrates of the formation reflect, to the degree they are able, the same temperate mild, moist climate as the plant fossils in sharp contrast to the present day arid conditions. The climate was somewhat cooler than earlier Clarno with possible freezing spells. A large diversity of carnivores including several cats as well as some mustelids and many species of dogs occurs here. Paleontologists in the past 20 years have been successful in turning up faunas of rabbits and rodents including squirrels, rats, mice and the larger beavers. Large populations of ungulates (hoofed mammals) are found here which only vaguely resemble present day relatives. These include tapirs, camels, horses, peccaries, pigs, and pig like artiodactyls as well as rhinoceros of both the aquatic and running or plains dwelling types. Over 100 genera of fossil vertebrates have been discovered in the John Day Formation, but the animals for which the John Day is best known are the oreodonts. These animals are artiodactyls (even-toed ungulates) and their particularly rich fossil record suggests large herds of the beasts. Oreodonts range in size from a dog up to a small horse. Their dentition clearly indicates that most were browsing herbivores. Placental mammals dominate the John Day, but one opossum (marsupial) has been reported (Stock and Furlong, 1922). Fragmentary material of both the upper and lower jaw are preserved, and the opossum *Peratherium* was similar in size to the modern Virginia opossum, *Didelphis*.

It has been speculated (Rensberger, 1972) that the basic drainage of the John Day area from a single large basin is little changed today from its topographic configuration during the Oligocene and Miocene epochs. Certainly the present day limits of the fossiliferous units would tend to support this conclusion. During John Day time on several occasions the area was interrupted by volcanic events with pronounced effects. In addition to ash beds of widespread distribution, there is evidence that ignimbrites were emplaced here in late Turtle Cove time. An ignimbrite results from the deposition of incandescent volcanic ash and related debris from an explosive volcano covering the landscape and cooling to a congealed mass. In spite of the pervasiveness and catastrophic nature of such an event, widespread animal and plant mortality is not evident in the fossil record; it must have occurred, nonetheless. Elsewhere today in the temperate and tropical zones where moist climate is extant, the scars of volcanic events are quickly healed and covered by plant activity and chemical weathering.

POST JOHN DAY VERTEBRATES

Vertebrate faunas of Barstovian (middle Miocene) age are found in several eastern Oregon localities as well as in the John Day Valley. Assemblages described from four major localities, Picture Gorge/Mascall Ranch area, Succor Creek, Skull Springs and Beatys Butte, are evidently synchronous. Barstovian vertebrate assemblages including primarily ungulates, carnivora and rodents are to be found at the Quartz Basin and Red Basin localities (Shotwell, 1968a, 1968b).

Downs (1956) described faunas from the Mascall Formation in some detail. Material he collected included a large number of carnivores and ungulates (hoofed mammals) including several camels, some elephants and deer. In addition, several rodents including the horned gopher *Mylagaulus* were reported. Several superb specimens of the Barstovian horse *Merychippus* show the range of natural variation of this form with respect to size and sexual variation.

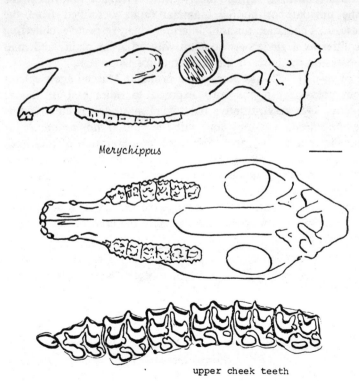

Merychippus ⎯⎯⎯

upper cheek teeth

223

The collections of large numbers of bones in discrete pockets of the Mascall Formation are interpreted by some authors as evidence of catastrophes in the geologic past such as dust storms or natural poisoning of the drinking waters. Geologically, the Mascall is presently interpreted as lacustrine (lake) and aeolian (windblown) tuffaceous sediments. A lake in the Mascall Ranch and Schneider Ranch areas may have resulted from blockage of stream drainage by volcanic activity. Downs (1956) and Chaney (1948) both note that the concentration of bones and leaves respectively in the Mascall is not nearly as high as the concentration is in the underlying John Day. This is interpreted by both authors as natural post mortem dispersal by a geologic agent such as a lake. The Mascall is believed to have been deposited in a valley with adjacent uplands including small lakes and streams with some swamps, bogs and wooded areas. The climate is estimated to have been moderate with more rain than today, and volcanic tuffs of the Mascall are dated radiometrically (Evernden, et al., 1964) at 15.4 m.y.b.p.

Vertebrate faunas from Succor Creek (Scharf, 1935), Skull Springs (Gazin, 1932) and Beatys Butte (Wallace, 1946) bear the same ungulate or hoofed mammal fauna described from the Mascall. The latter authors interpret their respective collecting localities as a glade or woodland with adjacent plains and mild climates on the basis of the vertebrate fossils present.

Downs (1952) has described a probable Mascall age elephant from volcanic ash and tuffs exposed 15 miles east of Baker, Oregon. This very primitive Barstovian mastodon is identified as *Gomphotherium*, synonymous with the genus *Trilophodon*, and is characterized by the triple loph (cusp) configuration of the cheek teeth. The most striking feature of this elephant, however, is the forward projecting elongate incisor teeth. The extension of the entire lower jaw is pronounced and the group is referred to as the "long jawed mastadons". It is suggested that the jaw appendage and teeth were for pulling up vegetable material or grubbing in forested uplands.

Gomphotherium

Another interesting report was of a Megalonychidae or sloth (Sinclair, 1906) from the Mascall Miocene. If this is correct, it is one of the earliest gravigrade remains in North America.

Quartz Basin and Red Basin Barsovian faunas in southeast Oregon are described by Shotwell (1968a). These include a large and diverse assemblage of insectivores and rodents as well as carnivores and ungulates. Some 63 species were reported from both areas and Shotwell concluded that the two faunas were of nearly identical age. Hutchinson (1966) described several species of shrews (insectivores) from the Quartz Basin as well as from the Skull Springs fauna.

Summarizing the Talpidae insectivores (moles), Hutchinson (1968) notes five separate "adaptive grades" for Oregon fossil and modern moles including ambulatory, aquatic, semiaquatic, semifossorial and fossorial. Several genera and species of insectivores including shrews and moles are reported and described from Barstovian sediments at Guano Lake about 40 miles east of Lakeview in Lake County.

Fry (1973) has described two Barstovian age turtles from the Alkali Canyon quarry where fossiliferous tuffaceous sediments are overlain by The Dalles Formation south and west of Arlington. These are accompanied by faunas of fossil ungulates and some rabbits. Fry notes the large size of these turtles which are nearly three feet long and suggests that they may have disappeared from Oregon as a result of climatic cooling.

Clarendonian stage (Miocene) faunas are described by Shotwell (1970) and Shotwell and Russell (1963) from the Black Butte locality in the Juntura Formation of the Juntura Basin. These are radiometrically dated at 11.3 m.y.b.p. and include an as-

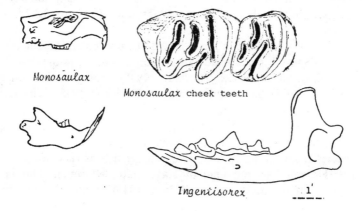

Monosaulax

Monosaulax cheek teeth

Ingeniisorex   1'

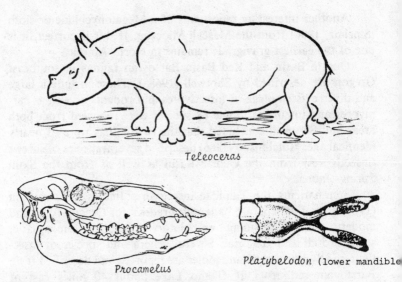

Teleoceras

Procamelus

Platybelodon (lower mandible)

semblage of insectivores (shrews) and rodents (mostly mice) with lesser numbers of carnivores and ungulates. *Proboscidea* or elephants including the genera *Mammut* and the "shovel tusker", *Platybelodon*, are represented here by some three separate species at two localities. The diversity of this assemblage suggests that it was probably a stream-bank environment.

The rhinoceros genus *Aphelops* reported by Shotwell (1963) from the Juntura Basin is a stratigraphically long ranging form that occurs from John Day time through the Pleistocene. Species of the rhino, *Teleoceras*, are characterized by a large body carried on short powerful legs similar in proportions to a hippopotamus with either no horn or only a short horn. Two genera of camel *Procamelus* and *Megatylopus* (*Paracamelus*) as well as a horse (*Hipparion*) are also abundant in the Black Butte fauna of the Juntura Basin.

The Rome fauna in southeastern Oregon, dated radiometrically at 10.5 m.y.b.p., has had various portions of the fauna described including the peccaries (Colbert, 1938), rodents and rabbits (Wilson, 1937,1938), and antelopes (Furlong, 1932). The latter author also noted a rhinoceros, horse, peccary, camel, dog and horned rodent from the same fauna as well as bears, a mole, otter and fish. Repenning (1967) in a monograph on shrews lists a new species from the Rome area which he regards as representative of the Hemphillian mammalian stage (late Miocene). The large number of aquatic forms led Wilson (1937) to interpret the Rome fauna as a lacustrine or lake environment. A species of badger described by Hall (1944) from Rome equivalent beds exposed on Stinking Water Creek in Harney County includes a well-preserved skull and lower jaw fragment. Hall suggests that the very early distinction between New World and Old World badgers may have been due to their largely insectivorous diet. The lack of insects year round at boreal latitudes restricted the badgers to more southerly temperate latitudes away from the boreal migration routes.

Shotwell (1970) described faunas from the Drewsey Formation of Hemphillian (late Miocene) Age from four localities: Bartlett Mountain, Drinkwater Basin, Otis Basin and in the western Juntura Basin. These included some 20 species of rodents and 6 shrews along with 3 carnivores including a bear and lesser numbers of other ungulates such as rhinoceros, deer, horses and elephants including *Amebelodon*. Shotwell was able to correlate the Drewsey material with the Rome, Rattlesnake and McKay Reservoir local faunas. Elsewhere the Drewsey formation is radiometrically dated as 8.5-8.9 m.y.b.p.

Hemphillian Stage faunas are described by Shotwell (1970) from the Juniper Creek Canyon locality in Malheur County and in the Little Valley locality in the Chalk Butte Formation from the same vicinity. Three shrews, one rabbit and thirteen rodents including two beavers are reported along with five carnivores including a bear (*Indarctos*). Ungulates from this site include a horse, rhinoceros and camel. Shotwell suggests that the Little Valley material is slightly younger than the Drewsey Formation material. The obvious wear and fragmentary nature of the bones at one small site here have prompted Shotwell to designate that material as redeposited.

Four localities in the McKay Reservoir area, which include two localities at Krebs Ranch, one at Westend Blowout and one at Boardman, have been reported by Shotwell (1956, 1958b). The Hemphillian material here is so fossiliferous that he has used these localities as the basis of a paleoenvironmental analysis over a broad area. In his classic study, Shotwell was able to distinguish three major environments in the Hemphillian to show a "pond bank," "woodland," and "grassland". He was also able to distinguish between an allochthonous or transported fauna and an autochthonous or in situ fauna by the statistical frequency of the separate bones of skeletons. In-place faunas are represented by a large array of skeletal elements, whereas, transported material is poorly represented and often fragmental and worn. The fauna at

Model community organization of Great Basin Hemphillian mammals
(after Shotwell, 1958)

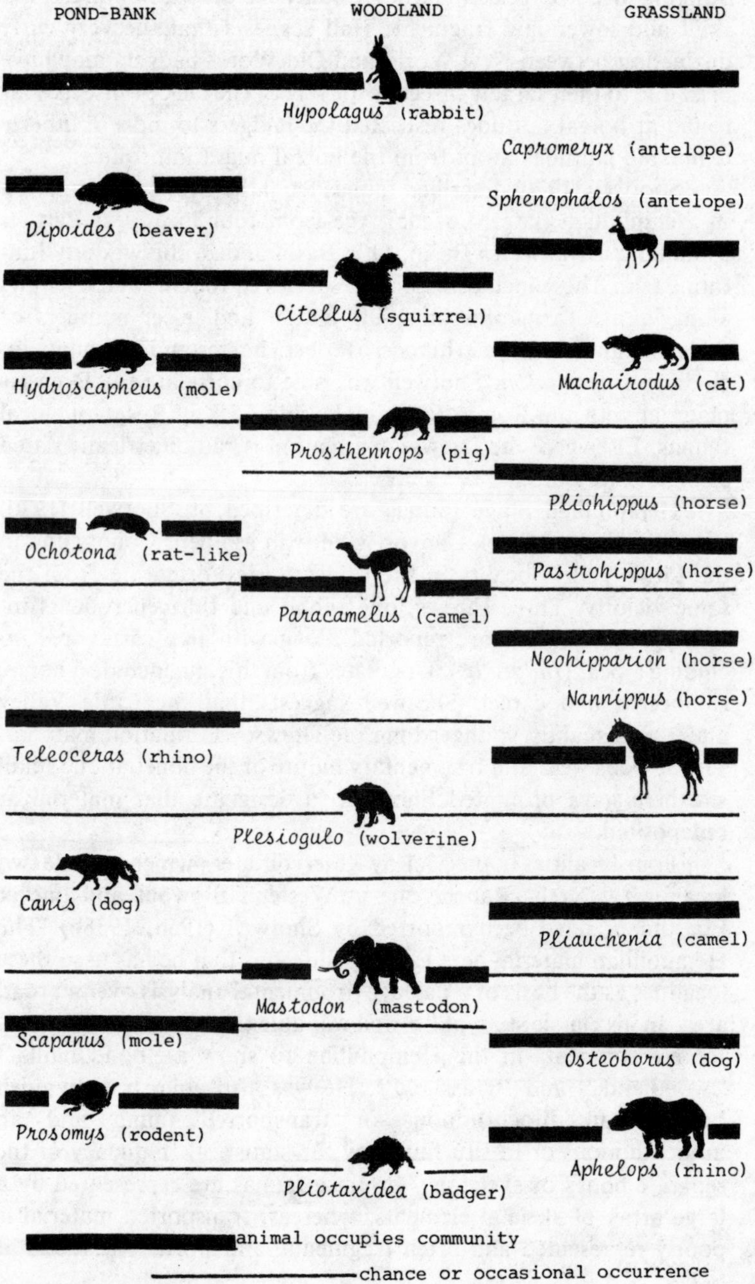

POND-BANK   WOODLAND   GRASSLAND

Hypolagus (rabbit)

Capromeryx (antelope)

Dipoides (beaver)   *Sphenophalos* (antelope)

Citellus (squirrel)

Hydroscapheus (mole)   *Machairodus* (cat)

Prosthennops (pig)

Pliohippus (horse)

Ochotona (rat-like)   *Pastrohippus* (horse)

Paracamelus (camel)   *Neohipparion* (horse)

Nannippus (horse)

Teleoceras (rhino)

Plesiogulo (wolverine)

Canis (dog)   *Pliauchenia* (camel)

Mastodon (mastodon)

Scapanus (mole)   *Osteoborus* (dog)

Prosomys (rodent)   *Aphelops* (rhino)

Pliotaxidea (badger)

▬▬▬▬▬▬ animal occupies community

──────── chance or occasional occurrence

228

these four synchronous localities is made up of shrews, rabbits, a bat, several rodents including two beavers, six carnivores including two dogs and two cats, an elephant, two horses, rhinoceros and camels as well as bones of frogs and turtle carapace fragments. The good condition of the McKay Reservoir material and the relative ease with which it is separated from the rich matrix of tuffaceous (ash) sediment contributed much to the success and accuracy of Shotwell's work. McKay Reservoir faunas bear a significant degree of similarity to Asiatic assemblages from North China suggesting animal migrations. The highest degrees of faunal similarity is in the carnivore group, whereas primates, edentates and artiodactyls of the two areas bear very little in common. Climatic evaluations for the McKay Reservoir material suggest that it was intermediate between the moist John Day time and the significantly drier Rattlesnake time.

Fossil plants of the McKay area show a gradual reduction of the Miocene hardwood forests, changing over to a conifer forest, then in the Pliocene/Pleistocene finally becoming a grasslands and subdesert environment. This succession through time in the late Tertiary is almost an exact reversal of the normal ecologic succession from a pioneer (grassland) stage through to a climax (hardwood forest) community. The reversal here is due, of course, to the development of the Cascade Range in late Tertiary time and to its subsequent rainshadow.

COMMUNITY INTERPRETATION OF FOUR HEMPHILLIAN FOSSIL LOCALITIES

Krebs 1 POND-BANK	Krebs 2 DISPLACED?	Westend Blowout WOODLAND	Boardman GRASSLAND

Hypolagus (rabbit)

Scapanus (mole)   Sphenophalos (antelope)

Dipoides (beaver)   Neohipparion (horse)

Citellus (squirrel)

Prosthennops (pig)

Pliauchenia (camel)

Paracamelus (camel)

Perognathus (rat)

Canis (dog)

Mastodon (mastodon)

Teleoceras (rhino)   Aphelops (rhino)

229

(after Shotwell, 1964)

The Rattlesnake Formation is separated from the underlying Mascall by a period of tilting and erosion producing an angular unconformity between the two formations. Consisting of water lain tuffs, silt and coarse gravel, the Rattlesnake has been radiometrically dated at 6.4 m.y.b.p. (Evernden, et al., 1964). In a monograph on North American squirrels (Black, 1963) a single species of *Citellus* (squirrel) is noted from the Rattlesnake Formation. A large edentate *Megalonyx* (sloth) is described from the Rattlesnake Formation in a monograph on Oregon fossil edentates. In that paper by Packard (1952) seventeen separate sloth discoveries in Oregon are noted including five from Fossil Lake and five from the Willamette Valley. The specimen from the Rattlesnake Formation is one of the few pre-Pleistocene sloths described. Thorpe (1922) described a dog-like *Araeocyon* carnivore from the Rattlesnake. The similarity of this form to Old World types suggests that it may be an intercontinental migrant. A large bear (*Indarctos*) similar to *I. oregonesis*, the size of a grizzly, was described (Merriam and Stock, 1927) from the same formation. The fossil material consists only of three upper left cheek teeth in a jaw fragment. What is more remarkable is the fact that ten years lapsed between 1916 when Stock and Moody collected part of the tooth row and 1926 when Merriam collected the remainder at the same locality.

The Rattlesnake fauna to date includes several genera (Merriam, Stock and Moody, 1925) of dogs, two bears, two mustelids, two cats, mice, two rabbits, beaver, two camels of a particularly large size, a peccary, a sloth, two horses, rhinos and a pronghorn antelope. Much of the fauna is modern in aspect, but significant portions of it, specifically the rhinoceros, elephants, and camels, relate it to the Tertiary.

There is an absence of browsing horses (Stock, 1946) in the Rattlesnake fauna, and the fossil record shows a progressive body size increase and toe reduction in the same group. This would corroborate Axelrod's (1966a) suggested increase in the grassland environment compared to the underlying forested Mascall. The presence of camels, mastodons, peccaries and rhinoceros, on the other hand, suggests some browsing habitats. Shotwell (1961) has demonstrated a sharp increase of the single-toed grazing horse *Pliohippus* in the Hemphillian that coincides with a diminished occurrence of the three-toed horse *Hipparion*. He was able to associate this faunal replacement with the simultaneous floral change in the Northern Great Basin from savanna to a grassland environment. Fossil leaves from the Rattlesnake Formation also imply a climatic picture significantly more dry than that of the John Day and Mascall Formations. Tuffaceous lacustrine sediments exposed three miles west of the town of Unity have yielded a variety of late Miocene (Hemphillian) mammals including a rhino (*?Peraceras*), horse (*Nannippus*), camel, sloth (Megalonchyloid) as well as elephants (*Trilophodon*) and a particularly well preserved skull of *Miomastodon*. Also present in these sediments are wood fragments, leaves and diatoms. Lowry (1943) refers to this unit informally as the "Ironside Formation," but his thesis remains unpublished.

Clear Pliocene faunas are not well established in Oregon. A recent restructuring of the Cenozoic time scale by Berggren and Van Couvering (1974) has redesignated much of the previous Pliocene to the late Miocene. A single mole, *Scapanus*, (Hutchinson, 1968) is recorded from possible Pliocene sediments at the Enrico Ranch locality near the southern Oregon border. Because most moles are "fossorial" or digging mammals, parts of the anatomy related to its fossorial habitat are extremely distinctive. The humerus bone in particular is enlarged and flattened as a support for the powerful muscles. This Enrico Ranch specimen is dated only as "at least as old as Blancan" (Pliocene) (Hutchinson, 1968).

Fossil shrew material, consisting of jaw and teeth parts, is reported from Christmas Valley late Hemphillian stage (Pliocene) (Repenning, 1967). Elsewhere Christmas Valley sediments are regarded as Pleistocene or "Late Pleistocene" (Allison, 1966). A new species of fossil turtle, *Clemmys owyheensis*, of "Pliocene Hemphillian" age was reported from near Rome on Dry Creek, Malheur County (Brattstrom, 1959). In view of the known Miocene Hemphillian mammalian vertebrate remains from this same vicinity and the poorly known chronology of fossil turtles, it seems likely that this locality is lower Hemphillian (late Miocene) and not Pliocene.

In a discussion of the Yonna Formation of the Klamath River Basin, the fossil bearing sediments are regarded as middle Pliocene. A small and indeterminate fauna of freshwater molluscs as well as freshwater diatoms from the same unit are dated as Pliocene by Ten-chien Yen and Lohman (in Newcomb, 1958). Probably the best evidence for the Yonna Formation Pliocene age is a peccary skull from the sediments exposed in "Wilson's Quarry Pit" identified as *Prosthennops oregonensis*. This form is considered "middle Pliocene," but no mammalian stage was designated.

Although Pleistocene mammals appear to be scattered over the

State, much of the bulk of the fossil material is concentrated in two areas. Thomas Condon first drew attention to Pleistocene deposits in the Willamette Valley when he referred to the valley as "Willamette Sound" (Condon, 1902). By studying sedimentary terraces, Condon estimated that the depth of water at Salem during Pleistocene time was as deep as 165 feet and as high as 325 feet over places in Portland. According to Condon, the Sound may have extended as far south as Eugene with a configuration not unlike Puget Sound of Washington.

Within the Valley, elephants of several species have been uncovered from bog deposits where they are often well preserved. An excellent summary of Pleistocene finds of mammals in Oregon is to be found in Hay (1927). Ground sloths of three separate genera have been collected at Fossil Lake as well as near Roseburg, Champoeg Creek in Clackamas County, near Eugene, and at Palmer Creek near Dayton in Yamhill County. Two of these sloths, *Nothrotherium* and *Megalonyx*, are roughly equivalent in size, being as big as a large ox. The largest Oregon sloth is *Mylodon* which was half again as large as the genera above. These sloths survived very late in the fossil record and there is good evidence they co-existed for a time in the later Pleistocene with man before disappearing. The largest North American sloth *Megatherium* at 20 feet in length and larger than an elephant is unreported from Oregon. Mastodons are known from Rye Valley in Baker county and bones were recovered from clays interbedded with gold bearing gravels near Arlington in Gilliam County, near Oregon City in Clackamas County, Rainier in Columbia County, near McMinnville in Yamhill County, on Nye Creek in Lincoln County, at Ten Mile Creek in Lane County, and at Myrtle Point, Coos County. Most of these mastodons are represented by the bunadont (low crowned) multicusp molars typical of the genus.

Nothrotherium

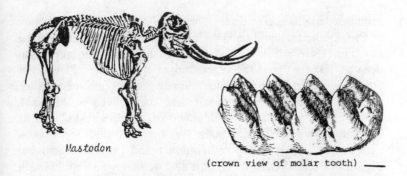

Mastodon

(crown view of molar tooth) ___

Elephants of the species, *Elephas boreus*, have been recorded in Oregon from near The Dalles. Remains of *Elephas columbi* have been collected from several sites east and west of the Cascades (Hay, 1927). The teeth and bones of Pleistocene bison and camel (*Camelops*) have been recovered from excavations in the vicinity of Portland and Oregon City. An excellent bison skull was reported by Condon (1902) from Pleistocene sediments near The Dalles. On the southern Oregon coast in Curry County a tapir tooth of the genus *Tapirus* sp. was collected in 1913 by J. C. Merriam (Leffler, 1964). The specimen was recovered from the late Pliocene/early Pleistocene Upper Elk River Formation where the accompanying molluscan fauna suggests a shallow offshore cold-water environment.

The second major Pleistocene fossil concentration in Oregon is in south central Oregon. Here several large dry lakes, including Fossil Lake, Christmas Lake and Silver Lake yield bones of vertebrates of all sizes. Again Thomas Condon played an important role in relaying fossil bones from these sites to paleontologist E. D. Cope of the Philadelphia Academy of Sciences. Both the Willamette Valley and south central Oregon sites are optimum for fossil preservation since both are basins with quiet shallow lakes and adjacent bogs conducive to the miring, entrapment, and preservation of large vertebrates. One such locality was described by Hansen and Packard (1949) wherein an early Holocene bog in the Silverton area yielded elephant bones and pollen. Vertebra, limb bones and a partial skull of the genus *Parelephas* were discovered in a swamp as it was being excavated to clear a spring near Evans Creek 3 miles east of Silverton. Although the entire elephant skeleton was probably present, it was never fully recovered, and bone specimens including both tusks were collected by construction workers and stored in a nearby barn where they soon dried and began to crack. The greatest variety and abundance of Pleistocene mammal remains in the State are to be found in the Fossil Lake site (Allison, 1966). Water in the prehistoric lake was as deep as 100 feet, and the vegetation of the adjacent Pleistocene landmass reflects a much wetter climate than occurs there today. Here, scattered among the beach sands and gravels of the basin, are bones and teeth of all manner of mammals, but mostly ungulates or hoofed mammals. Well represented are horses, camels, sloths, a peccary (*Platygonus*) and proboscideans, but Pleistocene bird skeletal material is also well known (Cope, 1889c). Associated with these same lake and stream sediments are aquatic mammals such as beavers. This includes the giant beaver, *Castoroides* (Matthew, 1902). This animal attained a length of over seven feet and grew to the size of a bear. Incisor teeth as large as bananas were operated by powerful jaw muscles.

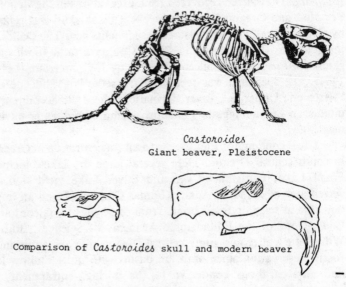

Castoroides
Giant beaver, Pleistocene

Comparison of *Castoroides* skull and modern beaver

South of Fossil Lake the horns and skull of a prehistoric big-horn sheep, *Ovis catclawensis*, have been recovered from Lake County near Adel (Thoms and Smith, 1973). Possible cultural material in the form of a battered worked pebble recovered with the skull suggest the skull may be even younger than Pleistocene.

As noted elsewhere in the text, the presence of fossils of the andromodous King Salmon, *Oncorhynchus tshawytscha*, in the vicinity of Klamath Falls (Jordan, 1907) suggests that the basin and

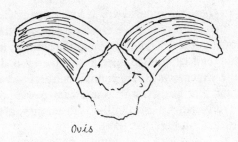

Ovis

range province may have emptied into the Pacific Ocean through the Klamath River as late as Pleistocene time.

Widespread extinctions have long attracted the attention of paleontologists, and one of the most profound of these occurs in Pleistocene faunas. Near the end of this epoch many of the large mammal species disappear from North America. These groups, including horses, camels, mammoth, mastodons, large bison, and various carnivores flourished throughout much of the late Tertiary as well as early Pleistocene only to vanish abruptly by the beginning of the Holocene epoch around 11,000 years ago. Attributing this large scale demise to the ice age conditions of the Pleistocene seems unreasonable as many of the species survived several previous ice advances of greater severity only to later disappear in North America near the end of the last ice withdrawal. The extinctions may correspond to the appearance of man as he migrated down from the Bering Strait area about this time and began intensively hunting out these large mammals. It has been suggested that

Bison

populations of these mammals were at low ebb anyway due to the ice ages and that man may have merely destroyed the remaining survivors. Some authors have postulated that hotter, more humid, climates at the end of the ice ages may have even driven larger animals to the waterways where human cultures were located. It should be noted that burros, horses and camels were reintroduced from Europe to North America in the 17th century, after their late Pleistocene extinction here.

The following classification contains Oregon fossil land
vertebrate genera listed in the published literature.
Taxonomic revision has not been attempted.

CLASS: Amphibia
  ORDER: Caudata
    FAMILY: Salamantridae (salamander)
      GENERA: *Palaeotaricha*
  ORDER: Anura
    FAMILY: Ranidae (frog)
      GENERA: *Ranid*
CLASS: Reptilia
  ORDER: Chelonia
    FAMILY: Testudinidae (turtle)
      GENERA: *Clemmys, Geochelone, Stylemys*
  ORDER:Squamata
    FAMILY Amphisbaenidae (lizard)
      GENERA: *Dyticonastis*
    FAMILY: Boidae (snake)
      GENERA: *Ogmophis*
  ORDER: Crocodilia (crocodile)
CLASS: Mammalia
  ORDER: Marsupialia
    FAMILY: Didelphidae (opossum)
      GENERA *Didelphid, Peratherium*
  ORDER: Chiroptera (bat)
  ORDER: Insectivora (insect eater)
    FAMILY: Erinaceidae (hedgehog)
      GENERA: *Lantanotherium, Micropternodus*
    FAMILY: Plesiosoricidae
      GENERA: *Meterix*
    FAMILY: Soricidae (shrew)
      GENERA: *Alluvisorex, Cryptotis, Hesperosorex,*
              Ingentisorex, Limnoecus, Mystipterus,
              Paracryptotis, Paradomnina, Trimylus (Het-
              erosorex)
    FAMILY: Talpidae (mole)
      GENERA: *Achlyoscapter, Domninoides, Gaillardia,*
              Hydroscapheus, Neurotrichus, Scalopoides,
              Scapanoscapter, Scapanus

  ORDER: Creodonta
    FAMILY: Hyaenodontidae (extinct carnivorous placentals)
      GENERA: *Hemipsalodon, Hyaenodon*
  ORDER: Carnivora (meat eaters)
    FAMILY: Canidae (dogs, wolves, fox)
      GENERA: *Aelurodon, Amphicyon, Araeocyon, Canis,*
              Cynodictis, Enhydrocyon, Euoplocyon, Gale-
              cynus, Hyaenocyon, Hypotemnodon, Mesocyon,
              Nothocyon, Osteoborus, Paradaphaenus, Parictis,
              Pericyon, Philotrox, Pliocyon, Temnocyon,
              Tephrocyon, Tomarctus, Vulpes.

CLASS: Mammalia
  ORDER: Carnivora (meat eaters)
    FAMILY: Felidae (cat)
      GENERA: *Archaelurus, Dinaelurus, Dinictis, Felis,
        Hoplophoneus, Machairodont, Machairodus,
        Nimravus, Pogonodon, Pseudaelurus*
    FAMILY: Mustelidae (weasel, mink, otter, skunk, badger)
      GENERA: *Brachypsalis, Eomellivora, Leptarctus, Lutra,
        Lutravus, Martes, Mustela, Oligobunis,
        Plesiogulo, Plionictis, Pliotaxidea,
        Potamotherium, Spilogale, Sthenictis, Taxidea*
    FAMILY: Procyonidae (raccoon, panda)
      GENERA: *Allocyon, Bassariscus*
    FAMILY: Ursidae (bear)
      GENERA: *Agriotherium (Hyaenarctos), Arctodus, Hemicyon,
        Indarctos, Ursus*
  ORDER: Amblypoda
    FAMILY: Pantolambdidae (large hoofed mammals)
      GENERA: *Pantadont*
    FAMILY: Uinatheriidae (large hoofed mammals)
      GENERA: *Uintatherium*
  ORDER: Proboscidea
    FAMILY: Elephantidae (elephant)
      GENERA: *Elephas*
    FAMILY: Gomphotheriidae (large-jawed mastodon)
      GENERA: *Amebelodon, Gomphotherium, Platybelodon,
        Tetrabelodon*
    FAMILY: Mastodontidae (mammoth)
      GENERA: *Mammut, Mastodon, Pliomastodon*
  ORDER: Artiodactyla ("even-toed" hoofed mammals)
    FAMILY: Entelodontidae (Elotheridae; "giant pigs")
      GENERA: *Boochoerus, Daeodon, Elotherium*
    FAMILY: Suidae (Old World pigs)
      GENERA: *Palaeochoerus*
    FAMILY: Tayæsuidae (Dicotylidae; New World peccaries)
      GENERA: *Cynorca, Desmathyus, Dicotyles, Perchoerus,
        Platygonus, Prosthenops, Thinohyus*
    FAMILY: Anthracotheriidae (pig-like)
      GENERA: *Bothroidon (Hyopotamus), Heptacodon,
        Octacodon*
    FAMILY: Agriocheroidae (clawed oreodonts)
      GENERA: *Agriochoerus*
    FAMILY: Merycoidodontidae (Oreodontidae; hoofed oreodonts)
      GENERA: *Dayohyus, Desmatochoerus, Eporeodon, Hypsilops,
        Merycochoerus, Oreodon, Oreodontoides
        (Paroreodon), Promerycochoerus, Pseudogeneto-
        choerus, Superdesmatochoerus, Ticholeptus,
        Ustatochoerus.*
    FAMILY: Camelidae (camel and llama)
      GENERA: *Alticamelus, Auchina, Cameloides, Camelops,
        Eschatius, Megatylopus, Miolabis, Paracamelus,
        Paratylopus, Pliauchenia, Poebrotherium,
        Procamelus, Protolabis, Protomeryx (Gomphother-
        ium), Tanupolama*

CLASS: Mammalia
  ORDER: Artiodactyla ("even-toed" hoofed mammals)
    FAMILY: Hypertragulidae (Modern pecora)
      GENERA: *Allomeryx, Hypertragulus, Leptomeryx*
    FAMILY: Protoceratidae
      GENERA: *Protoceras*
    FAMILY: Palaeomerycidae (primitive cervoids)
      GENERA: *Blastomeryx, Dromomeryx, Palaeomeryx,*
              Pediomeryx, Rakomeryx
    FAMILY: Cervidae (deer)
      GENERA: *Bourmeryx, Cervus, Odocoileus*
    FAMILY: Antilocapridae (early antelope)
      GENERA: *Antilocapra, Capromeryx, Ilingoceros,*
              Merycodus, Sphenophalos
    FAMILY: Bovidae (cow, sheep, goats)
      GENERA: *Bison, Ovis*
  ORDER: Edentata
    FAMILY: Megalonychidae (ground sloth)
      GENERA: *Megalonyx*
    FAMILY: Megalotheriidae (ground sloth)
      GENERA: *Nothrotherium*
    FAMILY: Mylodontidae (ground sloth)
      GENERA: *Mylodon*
  ORDER: Perissodactyla ("odd-toed" hoofed mammals)
    FAMILY: Equidae (horse)
      GENERA: *Anchippus, Anchitherium, Archaeohippus, Epi-*
              hippus, Equus, Hipparion, Hippidion
              (Hippidium), Hypohippus, Merychippus,
              Mesohippus, Miohippus, Nannippus, Neohippar-
              ion, Parahippus, Pliohippus, Protohippus.
    FAMILY: Brontotheriidae
      GENERA: *Brontotherium*
    FAMILY:Chalicotheriidae (clawed mammal)
      GENERA: *Chalicotherium, Moropus*
    FAMILY: Helaletidae (Hyrachyiidae; running rhinoceros)
      GENERA: *Colodon, Hyrachyus*
    FAMILY: Lophiodontidae (tapir-like)
      GENERA: *Lophiodon*
    FAMILY: Tapiridae (tapir)
      GENERA: *Protapirus, Tapirus*
    FAMILY: Amynodontidae (large amphibious rhinoceros)
      GENERA: *Amynodon*
    FAMILY: Rhinocerotidae (rhinoceros)
      GENERA: *Aceratherium, Aphelops (Teleoceras), Caenopus,*
              Diceratherium, Peraceras, Rhinoceros
  ORDER: Rodentia (rodent)
    FAMILY: Paramyidae (very primitive rodent)
      GENERA: *Prosciurus*
    FAMILY: Mylagaulidae (horned rodent)
      GENERA: *Epigaulus, Meniscomys, Mesogaulus, Mylagaulodon,*
              Mylagaulus

CLASS: Mammalia
  ORDER: Rodentia (rodent)
    FAMILY: Apolodontidae ("mountain beaver")
      GENERA: *Allomys, Haplomys, Liodontia, Meniscomys, Sewelleladon, Tardontia*
    FAMILY: Sciuridae (squirrels, chipmunk, marmot)
      GENERA: *Ammospermophilus, Arctomyoides, Citellus (Otospermophilus), Eutamias, Marmota, Miosciurus, Protosciurus, Protospermophilus, Sciuropterus, Sciurus, Spermophilus*
    FAMILY: Ochotonidae
      GENERA: *Hesperolagomys, Ochotona, Oreolagus*
    FAMILY: Cricetidae (mice, muskrat)
      GENERA: *Arvicola, Hesperomys, Microtoscoptes (Goniodontomys), Microtus, Mimomys, Odatra, Oryzomys, Paciculus, Peromyscus, Prosomys, Synaptomys*
    FAMILY: Zapodidae (jumping mice)
      GENERA: *Macrognathomys, Pliozapus*
    FAMILY: Eomyidae (chinchillas, spiny rats, guinea pigs)
      GENERA: *Adjidaumo, Leptodontomys, Pseudotheridomys*
    FAMILY: Geomyidae (pocket gopher)
      GENERA: *Entoptychus, Geomys, Pleurolicus, Pliosaccomys, Schizodontomys, Thomomys*
    FAMILY: Heteromyidae (rats)
      GENERA: *Cupidinomys, Dipodomys, Diprionomys, Parapliosaccomys, Peridiomys, Perognathus, Prodipodomys, Proheteromys*
    FAMILY: Castoridae (beaver)
      GENERA: *Castor, Castoroides, Chalicomys, Dipoides, Eucastor, Hystricops, Monosaulax, Stenofiber*
  ORDER: Lagomorpha
    FAMILY: Leporidae (rabbits, hare, pika)
      GENERA: *Hypolagus, Lepus, Palaeolagus*

OREGON VERTEBRATE FOSSIL BEARING FORMATIONS

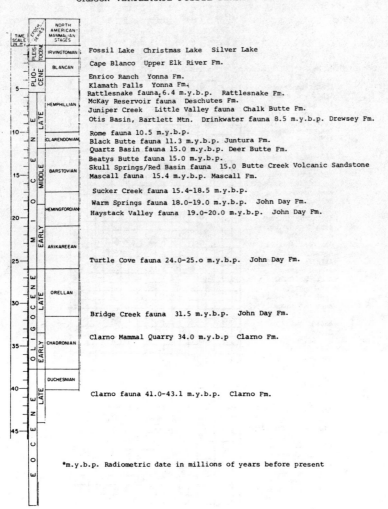

TIME SCALE IN B.P.	EPOCH SERIES	NORTH AMERICAN MAMMALIAN STAGES	

Fossil Lake   Christmas Lake   Silver Lake

Cape Blanco   Upper Elk River Fm.

Enrico Ranch   Yonna Fm.
Klamath Falls   Yonna Fm.
Rattlesnake fauna 6.4 m.y.b.p.   Rattlesnake Fm.
McKay Reservoir fauna   Deschutes Fm.
Juniper Creek   Little Valley fauna   Chalk Butte Fm.
Otis Basin, Bartlett Mtn.   Drinkwater fauna 8.5 m.y.b.p.   Drewsey Fm.

Rome fauna 10.5 m.y.b.p.
Black Butte fauna 11.3 m.y.b.p.   Juntura Fm.
Quartz Basin fauna 15.0 m.y.b.p.   Deer Butte Fm.
Beatys Butte fauna 15.0 m.y.b.p.
Skull Springs/Red Basin fauna   15.0 Butte Creek Volcanic Sandstone
Mascall fauna   15.4 m.y.b.p.   Mascall Fm.

Sucker Creek fauna 15.4-18.5 m.y.b.p.

Warm Springs fauna 18.0-19.0 m.y.b.p.   John Day Fm.

Haystack Valley fauna   19.0-20.0 m.y.b.p.   John Day Fm.

Turtle Cove fauna 24.0-25.o m.y.b.p.   John Day Fm.

Bridge Creek fauna   31.5 m.y.b.p.   John Day Fm.

Clarno Mammal Quarry 34.0 m.y.b.p   Clarno Fm.

Clarno fauna 41.0-43.1 m.y.b.p.   Clarno Fm.

*m.y.b.p. Radiometric date in millions of years before present

OREGON FOSSIL VERTEBRATE RANGES

The following are fossil vertebrate ranges listed in the published literature.

	Clarno Fm.	Clarno Mammal Quarry	Bridge Creek fauna	Turtle Cove fauna	Haystack Valley fauna	Warm Springs fauna	Sucker Creek Fm.	Mascall Fm.	Red Bsn.,Skull Sp.	Quartz Basin fauna	Beatys Butte fauna	Black Butte fauna	Rome fauna	Drinkwr., Otis, Bart. Mt	McKay, Ltle.Vly, Junip.Cr	Rattlesnake Fm.	Yonna Fm.	Enrico Ranch fauna	Elk River	Pleistocene
Aceratherium (rhino)				●					●											
Achlyoscapter (mole)								●												
Adjidaumo (rodent)								●												
Aelurodon (dog)														●	●					
Agriochoerus			●	●																●
Agriotherium (bear)																				●
Allocyon			●																	
Allomeryx			●																	
Allomys (mtn. bvr)			●	●																
Alluvisorex (shrew)									●	●				●						
Alticamelus (camel)								●						●						
Amebelodon (mastodon)															●					
Ammospermophilus													●							
Amphicyon (dog)							●	●	●							●				
Amynodon (rhino)	●																			
Anchippus (horse)			●				●													
Anchitherium (horse)			●	●																
Antilocapra																				●
Aphelops (rhino)				●				●					●	●	●	●				●
Araeocyon (dog)																●				
Arcaelurus (cat)			●	●																
Archaeohippus					●			●												
Arctodus (bear)																				●
Arctomyoides								●												
Arvicola (rodent)																				●
Auchina (camel)																				●
Bassarisus											●									●
Bison																				●
Boochoerus (pig)			●	●																
Bothroidon				●																
Bouromeryx (deer)				●																
Blastomeryx (cervoid)				●	●		●													
Brachypsalis (mustelid)								●	●											
Brontotherium	●																			
Caenopus (rhino)			●																	
Camelops (camel)																				●
Canis (dog)								●	●	●	●	●	●		●	●		●		●
Capromeryx (antelope)															●					
Castor (beaver)				●											●	●				
Castoroides (beaver)																				●
Cervus (deer)								●												
Chalicomys (beaver)							●	●												
Chalicotherium							●	●												
Chiroptera (bat)		●												●		●				
Citellus (squirrel)										●		●		●		●	●			●
Clemmys (turtle)									●					●						
Colodon (rhino)			●																	
Crocodilia	●																			
Cryptotis (shrew)																				●
Cupidinomys (rodent)														●						
Cynodictis (dog)			●																	
Cynorca (peccary)				●																

242

Column headers (left to right): Clarno · Clarno Mam. · Br. Cr. · T. Cove · Haystk. · Warm Spr. · Sucker · Mascall · Red. Bsn. · Qtz. Bsn. · Beatys B. · Black Bu. · Rome · Drink. · Mckay · Rattle. · Yonna · Enrico · Elk River · Pleistoc.

Genus	
Daeodon (pig)	
Desmatochoerus	
Desmathyus (peccary)	
Dicotyles (peccary)	
Diceratherium (rhino)	
Didelphid (opossum)	
Dinaelurus (cat)	
Dinictis (rat)	
Dinohyus	
Dipodomys (rat)	
Dipoides (beaver)	
Diprionomys (rat)	
Domninoides (mole)	
Dyticonastis (lizard)	
Dromomeryx (cervoid)	
Elephas	
Elotherium (pig)	
Enhydrocyon (dog)	
Entoptychus (rodent)	
Eomellivora (mustelid)	
Epigaulus (rodent)	
Epihippus (horse)	
Eporeodon (oreo.)	
Equus (horse)	
Eschatius (camel)	
Eucastor (beaver)	
Euplocyon (dog)	
Eutamias (squirrel)	
Felis (cat)	
Gaillardia (mole)	
Galecynus (dog)	
Geochelone (turtle)	
Geomys (rodent)	
Gomphotherium	
Haplomys (mtn. bvr)	
Hemicyon (bear)	
Hemipsalodon	
Heptacodon	
Hesperomys (mouse)	
Hesperosorex	
Hesperolagomys	
Hipparion (horse)	
Hippidion (horse)	
Hoplophoneus (cat)	
Hyaenocyon (dog)	
Hyaenodon	
Hydroscapheus (mole)	
Hypertragulus (pecora)	
Hypohippus (horse)	
Hypolagus (rabbit)	
Hypotemnodon (dog)	
Hypsilops	
Hyrachyus (rhino)	
Hystricops (beaver)	
Ilingoceros (antelope)	
Indarctos (bear)	
Ingentisorex (shrew)	
Lantanotherium (hedgehog)	
Leptarctus (mustelid)	
Leptodontomys (spiny rat)	

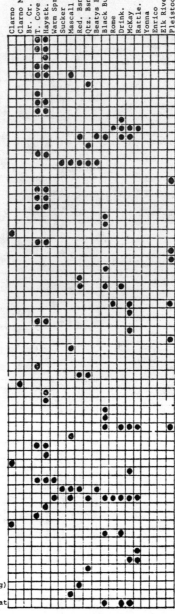

243

Genus	Clarno	Clarno Mam.	Br. Cr.	T. Cove	Haystk.	Warm Spr.	Sucker	Mascall	Red. Bsn.	Qtz. Bsn.	Beatys B.	Black Bu.	Rome	Drink.	McKay	Rattle.	Yonna	Enrico	Elk River	Pleistoc.
Leptomeryx (pecora)			●	●																●
Lepus (rabbit)			●	●																
Limnoecus (shrew)								●												
Liodontia ("mtn. beaver")											●	●		●	●					
Lophiodon	●																			
Lutra (mustelid)														●						●
Lutravus (mustelid)																				
Machairodus (cat)				●										●	●					
Macrognathomys (rodent)										●		●		●	●				●	
Mammut (mammoth)														●	●					
Marmota (squirrel)										●										
Martes (mustelid)									●			●								
Mastodon (mammoth)													●	●						●
Megalonyx (sloth)											●	●								
Megatylopus (camel)											●	●								
Meniscomys (rodent)			●	●																
Merychippus (horse)				●	●	●	●	●	●	●	●	●								
Merycochoerus				●	●															
Merycodus (antelope)					●	●	●	●	●											
Mesocyon (dog)			●	●																
Mesogaulus (rodent)				●	●															
Mesohippus (horse)				●																
Metamynodon	●																			
Meterix								●	●	●										
Micropternodus				●																
Microtoscoptes (mouse)											●	●	●	●						
Microtus (mouse)																				●
Mimomys (mouse)																				●
Miohippus (horse)			●	●	●															
Miolabis (camel)				●	●															
Miosciurus (squirrel)									●	●										
Monosaulax (beaver)									●	●										
Moropus	●		●	●			○													
Mustela (mustelid)										●	●									
Mylagaulodon (rodent)			●	●																
Mylagaulus (rodent)								●	●	●	●	●	●	●	●					
Mylodon (sloth)										●	●									
Mystipterus (shrew)									●	●	●									
Neohipparion (horse)														●	●					
Neurotrichus (mole)																				
Nimravus (cat)			●	●																
Nothocyon (dog)			●	●																
Nothrotherium (sloth)														●						●
Ochotona (rodent)			●	●																●
Octacodon			●	●																●
Odocoileus (cervid)																				●
Ogmophis (snake)				●	●															
Oligobunis (mustelid)				●	●															
Ondatra (rodent)																				●
Oreodon			●	●																
Oreodontoides			●	●																
Oreolagus (rodent)											●									
Oryzomys (rodent)																●	●			
Osteoborus (dog)															●	●				●
Ovis (sheep)																				
Paciculus (rodent)			●	●																
Palaeochoerus (pig)			●	●																
Palaeolagus (rabbit)			●	●	●															
Palaeomeryx (cervoid)						●														
Palaeotaricha (salamander)	✳																			
Pantodont														●						
Paracryptotis (shrew)																				

244

	Clarno	Clarno Mam.	Br. Cr.	T. Cove	Haystk.	Warm Spr.	Sucker	Mascall	Red. Bsn.	Qtz. Bsn.	Beatys B.	Black Bu.	Rome	Drink.	McKay	Rattle.	Yonna	Enrico	Elk River	Pleistoc.
Paradaphaenus (dog)				•	•															
Paradomnia (shrew)								•												
Parahippus (horse)					•	•	•	•												
Parapliosaccomys (rat)														•						
Paratylopus (camel)			•	•																
Parelephas (elephant)																				•
Parictis (dog)			•	•																
Pediomeryx (cervoid)														•						
Peraceras (rhino.)																				
Peratherium (opossum)			•	•																
Perchoerus (peccary)			•	•																
Pericyon (dog)			•	•																
Peridiomys (rat)							•	•	•											
Perognathus (rat)								•	•				•	•	•					
Peromyscus (mouse)					•			•	•	•			•	•	•					
Philotrox (dog)			•																	
Platybelodon (mastodon)								•												
Platygonus (peccary)									•				•		•					
Plesiogulo (mustelid)			•	•																
Pleurolicus (rodent)			•	•																
Pliauchenia (camel)																•				
Pliocyon (dog)																•				
Pliohippus (horse)													•	•	•					
Pliomastodon (mastodon)													•							
Plionictis (mustelid)													•							
Pliosaccomys (rodent)													•	•	•					
Pliotaxidea (mustelid)													•	•	•					
Pliozapus (rodent)														•						
Poebrotherium (camel)			•																	
Pogonodon (cat)			•	•																
Potamotherium (mustelid)						•														
Procamelus (camel)								•					•	•	•					
Prodipodomys (rat)								•	•	•										
Proheteromys (rat)								•												
Promerycochoerus (oreo.)			•	•	•															
Prosciurus (squirrel)			•																	
Prosthenops (peccary)					•	•	•	•			•	•	•	•	•					
Prosomys (rodent)															•					•
Protoceras (pecora)			•																	
Protohippus (horse)			•					•												
Protolabis (camel)			•																	
Protomeryx (camel)			•	•																
Protosciurus (squirrel)			•																	
Protospermophilus			•					•												
Protapirus (tapir)			•	•																
Pseudaelurus (cat)								•					•		•					
Pseudogenetochoerus			•	•																
Pseudotheridomys								•	•											
Rakomeryx (cervoid)								•												
Ranid (frog)																•				
Rhinoceros			•	•	•															
Scalopoides (mole)								•	•				•		•					
Scapanoscapter								•												
Scaphanus (mole)													•	•	•	•	•		•	
Sciuropterus								•												
Sciurus (squirrel)			•					•	•											
Sewelleladon			•																	
Spermophilus								•	•				•	•	•					
Sphenophalos													•	•	•					
Spilogale (mustelid)																				•
Sthenictis (mustelid)								•												

	Clarno Fm.	Clarno Mammal Quarry	Bridge Creek fauna	Turtle Cove fauna	Haystack Valley fauna	Warm Springs fauna	Sucker Creek Fm.	Mascall Fm.	Red Bsn., Skull Sp.	Quartz Basin fauna	Beatys Butte fauna	Black Butte fauna	Rome fauna	Drinkwr., Otis, Bart. Mtn.	McKay, Ltle.Vly, Junip.Cr.	Rattlesnake Fm.	Yonna Fm.	Enrico Ranch fauna	Elk River	Pleistocene
Stenofiber (beaver)			●	●																
Stylemys (turtle)		●	●	●																
Superdesmatochoerus				●																
Synaptomys (rodent)															●					
Tanupolama (camel)																				●
Tapirus (tapir)																			●	●
Tardontia (mtn. bvr)									●		●									
Taxidea (mustelid)																				●
Temnocyon (dog)			●	●																
Tephrocyon (dog)																				
Testudinidae (turtle)															●					
Tetrabelodon (mastod.)																●				
Thinohyus (peccary)		●	●	●																
Thomomys (rodent)																				●
Ticholeptus (oreo.)					●	●	●	●												
Tomarctus (dog)							●	●												
Trimylus (shrew)							●													
Uintatherium	●																			
Ursus (bear)																				●
Ustatochoerus														●	●					
Vulpes (dog)													●							●

Fisher Fm. ✳

Astoria Fm. ○

OREGON FOSSIL VERTEBRATE LOCALITIES

Following is a list of Oregon fossil vertebrate localities. The reference numbers are to authors listed in the bibliography who have published on the specific localities.

EOCENE
1. Clarno, Wheeler Co. Clarno Fm. Refs. 203, 171b.
2. Iron Mtn., 2 mi. E of Clarno, Wheeler Co. Clarno Fm.
     Refs. 71, 210.

EARLY OLIGOCENE
3. Clarno Mammal Quarry, 2 mi. E of Clarno, Wheeler Co.
     Refs. 128.

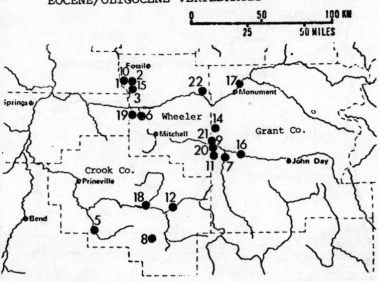

EOCENE/OLIGOCENE VERTEBRATES

LATE OLIGOCENE

4. Alamo Ranch (on John Day River), Grant Co. John Day Fm.
   Refs. 182. (Locality uncertain)
5. Bear Creek valley, Crook Co. John Day Fm. Refs. 113.
6. Bridge Creek valley, Wheeler Co. John Day Fm. Refs. 110,
   110a, 110b, 133, 182, 190b, 214a, 214e.
7. Camp Watson (on John Day River), Grant Co. John Day Fm.
   Refs. 214e.
8. Camp Creek valley, Crook Co. John Day Fm. Refs. 23, 182.
9. Cants Ranch, 10 mi. W of Dayville on John Day River, Grant
   Co. (Butler Basin). John Day Fm. Refs. 182.
10. Clarno, Wheeler Co. John Day Fm. Refs. 190b, 214, 214a,
    214e.
11. Cottonwood Creek valley, 4 mi. W of Dayville, Grant Co.
    John Day Fm. Refs. 122, 214, 214e.
12. Crooked River valley (on Post-Paulina Road), Crook Co.
    John Day Fm. Refs. 110a, 110b.
13. Foree Fossil Beds, Grant Co. John Day Fm. Refs. 60.
14. Goshen, Lane Co. ?Fisher Fm. Late Oligocene. Refs. 222.
15. Iron Mtn., Wheeler Co. John Day Fm. Refs. 171b
16. John Day basin (general locality), Grant Co. John Day
    Fm. Refs. 78a, 115, 122a, 182, 214b, 214a.
17. John Day River (north fork; general locality), Grant Co.
    John Day Fm. Refs. 41h, 133, 182, 214, 214a.
18. Logan Butte, 12 mi. S of Crooked River, Crook Co. John
    Day Fm. Refs. 18, 60, 132e, 206, 214a, 214e.
19. McAllister's Ranch, below Twickenham, Wheeler Co. John
    Day Fm. Refs. 190b.
20. Sheep Rock, Grant Co. John Day Fm. Refs. 60, 171b, 182.
21. Turtle Cove Fossil Beds, Grant Co. John Day Fm. Refs. 19,
    54, 60, 78a, 115, 133, 171b, 182, 190b, 214, 214a, 214e.
22. Spray, Wheeler Co. John Day Fm. Refs. 190.

247

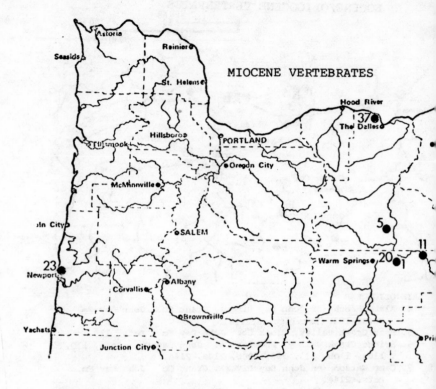

MIOCENE VERTEBRATES

EARLY MIOCENE FOSSIL VERTEBRATE LOCALITIES (HEMINGFORDIAN)

1. Gateway (SE of), Jefferson Co. John Day Fm. Refs. 81.
2. Haystack Creek valley, 5 mi. NE of Spray, Wheeler Co.
    John Day Fm. Refs. 60, 115, 133, 171, 171a, 171b, 182,
    185b, 190b, 214, 214a, 214e.
3. John Day basin (general locality), Wheeler Co. John Day
    Fm. Refs. 19, 41a-41f, 41j, 120, 120b, 122, 132a, 133.
4. Johnson Creek, SW of Kimberly, Wheeler Co. John Day Fm.
    Refs. 133, 185b, 190, 190a.
5. Mutton Mtns., N of Warm Springs, Wasco Co. John Day
    Fm. Refs. 244.
6. Picture Gorge, 5 mi. S of Kimberly, Grant Co. John Day
    Fm. Refs. 60, 171, 171a.
7. Powell's Ranch, 5 mi. above Ritter, Grant Co. John Day
    Fm. Refs. 190b.
8. Rudio Creek, 5 mi. SE of Kimberly, Grant Co. John Day
    Fm. Refs. 60, 171, 171a, 181, 190b, 204.
9. Schrock's Ranch, NE of Sheep Mtn. on Camp Creek, Crook Co.
    John Day Fm. Refs. 18, 60, 171, 171a.
10. Sutton Mtn., NW of Mitchell, Wheeler Co. John Day Fm.
    Refs. 171b.
11. Trout Creek valley, near junction with Antelope Creek,
    Jefferson Co. John Day Fm. Refs. 81.

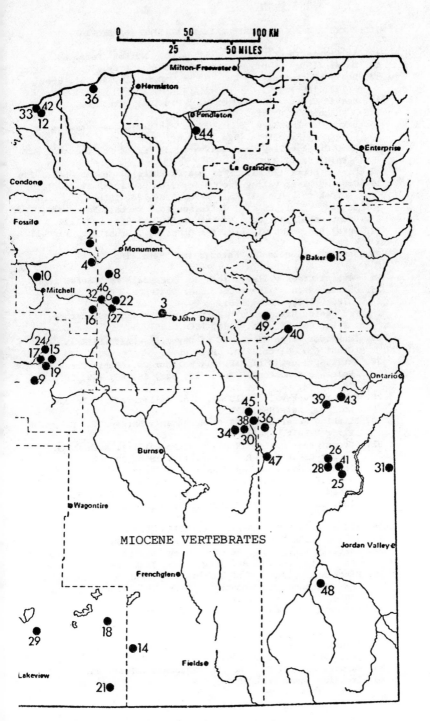

MIOCENE VERTEBRATES

MIDDLE MIOCENE FOSSIL VERTEBRATE LOCALITIES (BARSTOVIAN)

12. Arlington (4 mi. S of), Gilliam Co. No Fm. Refs. 64.
13. Baker (15 mi. E of), Baker Co. No Fm. Refs. 52.
14. Beatys Butte, Harney Co. No Fm. Refs. 19, 46, 52, 88a,
    172, 186b, 226.
15. Beaver Creek valley (3 mi W of Paulina), Crook Co.
    Mascall Fm. Refs. 52a, 78, 78a.
16. Birch Creek valley, 6 mi. SW of Picture Gorge, Wheeler Co.
    Mascall Fm. Refs. 52a.
17. Camp Creek valley (on Paulina road), Crook Co. Mascall Fm.
    Refs. 52a, 133.
18. Corral Butte (20 mi. NW of Beatys Butte). No Fm. Refs. 226.
19. Crooked River valley (south fork, on Paulina-Post road),
    Crook Co. Mascall Fm. Refs. 52a, 124, 186b.
20. Gateway (3 mi. SE of), Jefferson Co. Mascall Fm. Refs. 52a.
21. Guano Lake, Lake Co. No Fm. Refs. 88, 88a, 123, 172.
22. McKay Ranch (NE of Stewart's Crossing), Grant Co. Mascall
    Fm. Refs. 52a.
23. Newport, Lincoln Co. Astoria fm. Refs. 145.

24. Paulina Creek valley, Crook Co. ?Mascall Fm. Refs. 186b.
25. Quartz Basin, Malheur Co. Deer Butte Fm. Refs. 88, 88a,
    186e, 186g, 186h, 186i.
26. Red Basin, Malheur Co. Butte Creek Volcanic Sandstone.
    Refs. 66, 88, 88a, 186g, 186h, 186i.
27. Schneider Ranch (5 mi. NE of Dayville; White Hills),
    Grant Co. Mascall Fm. Refs. 52a.
28. Skull Springs (S of Harper), Malheur Co. Butte Creek
    Volcanic Sandstone. Refs. 19, 52a, 66, 88, 179, 186b,
    186i, 224.
29. Snyder Creek valley (10 mi. E of Valley Falls), Lake Co.
    No. Fm. Refs. 88a, 172.
30. Stinking Water Creek (NE of Buchanan), Harney Co. Columbia
    River Basalt. Refs. 70.
31. Succor Creek valley (9 mi. N of Rockville), Malheur Co.
    Sucker Creek Fm. Refs. 52a, 179, 186b, 206, 226.
32. Thomas Condon-John Day Fossil Beds State Park, Grant Co.
    (Mascall Ranch and area between Cottonwood and Rattlesnake
    Creeks). Mascall Fm. Refs. 19, 52a, 66, 122a, 132a,
    133, 134, 155e, 171b, 190c, 205a, 124e, 226.

LATE MIOCENE FOSSIL VERTEBRATE LOCALITIES (HEMPHILLIAN)
33. Arlington, Gilliam Co. No Fm. Refs. 19, 88.
34. Bartlett Mtn., 5 mi. SE of Drewsey, Harney Co. Drewsey
    Fm. Refs. 19, 70, 88, 186g, 186h, 186k, 187.
35. Black Butte, 3 mi. W of Juntura. Malheur Co. Juntura
    Fm. Miocene (Clarendonian). Refs. 19, 88, 186b, 186d,
    186g, 186h, 186i, 186k.
36. Boardman, Morrow Co. No Fm. Refs. 186c.
37. Chenoweth Creek valley, 3 mi. NW of The Dalles, Wasco Co.
    The Dalles Fm. Refs. 221.
38. Drinkwater Pass, 5 mi. E of Drewsey, Harney Co. Drewsey
    Fm. Refs. 19, 88, 186g, 186h, 186k, 187.
39. Harper, Malheur Co. No Fm. ?Miocene. Refs. 65.
40. Ironside (near Post Office), Malheur Co. "Ironside Fm."
    Refs. 113a, 132c.

41. Juniper Creek valley (30 mi. SE of Juntura), Malheur Co. No Fm. Refs. 186g, 186h, 186k.
42. Krebs Ranch, SE of Arlington, Gilliam Co. No Fm. Refs. 19, 88, 186b, 186c, 186g.
43. Little Valley (10 mi. SW of Vale), Malheur Co. Chalk Butte Fm. Refs. 88, 186g, 186h, 186k.
44. McKay Reservoir, Umatilla Co. ?Deschutes Fm. Refs. 19, 88, 186a, 186b, 186c, 186g, 186h, 172a.
45. Otis Basin (5 mi. NE of Drewsey), Harney Co. Drewsey Fm. Refs. 19, 88, 186e, 186g, 186h.
46. Picture Gorge area, Grant Co. Rattlesnake Fm. Refs. 19, 37, 78, 78a, 122a, 132d, 133, 134, 135, 155e, 214c, 124d, 241a.
47. Riverside, Malheur Co. No Fm. Miocene (Clarendonian). Refs. 88.
48. Rome (5 mi. SW of on Dry Creek), Malheur Co. No Fm. Refs. 24, 37, 65, 88, 172, 172a, 186b, 241, 241a.
49. Unity, Baker Co. "Ironside Fm." Refs. 113a.

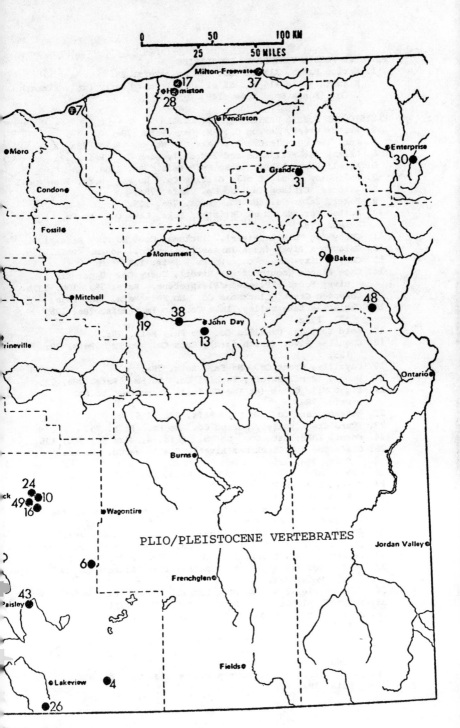

PLIO/PLEISTOCENE VERTEBRATES

PLIOCENE FOSSIL VERTEBRATE LOCALITIES
1. Enrico Ranch, Klamath Co. No Fm. Refs. 88.
2. Klamath Falls (15 mi. NE of at Wilson's Quarry Pit), Klamath Co. Yonna Fm. Refs. 148.

PLEISTOCENE FOSSIL VERTEBRATE LOCALITIES
3. Abiqua Creek, Marion Co. No Fm. Refs. 78a.
4. Adel (on 20 Mile Creek), Lake Co No Fm. Refs. 78a, 213.
5. Albany (at Mill Race and 2nd St.), Linn Co. Refs. 125.
6. Alkali Flat, Lake Co. No Fm. Refs. 78a, 110a.
7. Arlington (5 mi. E of), Gilliam Co. No Fm. Refs. 78a.
8. Astoria, Clatsop Co. No Fm. Refs. 78a.
9. Baker, Baker Co. No Fm. Refs. 78a, 125.
10. Button Springs, N of Christmas Lake, Lake Co. No Fm. Refs. 78a.
11. Canby (on Molalla River), Clackamas Co. No Fm. Refs. 125.
12. Calapooya River (near Tangent), Linn Co. No Fm. Refs. 125.
13. Canyon City, Grant Co. No Fm. Refs. 125.
14. Cape Blanco (mouth of Elk River), Curry Co. Upper Elk River Beds. ?Pliocene/Pleistocene. Refs. 78a, 109, 132b.
15. Champoeg Creek, Clackamas Co. No Fm. Refs. 78a, 155e.
16. Christmas Lake valley, Lake Co. No Fm. Refs. 78a, 125. 172, 172a.
17. Cold Springs, Umatilla Co. No Fm. Refs. 78a.
18. Coquille River (north fork), Coos Co. No Fm. Refs. 78a, 125.
19. Dayville, Grant Co. No Fm. Refs. 78a.
20. Dundee (2 mi. SW of), Yamhill Co. No Fm. Refs. 78a, 125.
21. Eight Mile Creek (at The Dalles), Wasco Co. No Fm. Refs. 78a.
22. Eugene, Lane Co. No Fm. Refs. 78a, 125.
23. Evans Creek valley, Marion Co. No Fm. Refs. 75.
24. Fossil Lake, Lake Co. No Fm. Refs. 4, 41m, 54, 78a, 125.
25. Gladstone (on Clackamas River), Clackamas Co. No Fm. Refs. 78a, 125.
26. Goose Lake, Lake Co. No Fm. Refs. 78a, 125.
27. Harrisburg (on Willamette River), Linn Co. No Fm. Refs. 78a, 125.
28. Hermiston (3 mi. E of), Umatilla Co. No Fm. Refs. 78a.
29. Jacksonville, Jackson Co. No Fm. Refs. 78a, 125.
30. Joseph (10 mi. E of), Wallowa Co. No Fm. Refs. 78a.
31. LaGrande (2 mi. SE of), Union Co. No Fm. Refs. 78a, 165.
32. Lebanon (4 mi. E of), Linn Co. No Fm. Refs. 78a, 125.
33. Lost River (15 mi. E of Klamath Falls), Klamath Co No Fm. Refs. 78a.
34. McMinnville (2 mi. NW of), Yamhill Co. No Fm. Refs. 78a.
35. Merrill, Klamath Co. No Fm. Refs. 78a, 155f.
36. Mill Creek (near Woodburn), Marion Co. No Fm. Refs. 78a, 155f.
37. Milton-Freewater, Umatilla Co. No Fm. Refs. 78a.
38. Mt. Vernon (between Mt. Vernon and Canyon City on John Day River). No Fm. Refs. 78a, 125.
39. Myrtle Creek (SE of Roseburg), Douglas Co. No. Fm. Refs. 78a, 125.

40. Newberg, Yamhill Co. No Fm. Refs. 78a, 125.
41. Nye Creek (at mouth of), Lincoln Co. No Fm. Refs. 78a.
42. Oregon City, Clackamas Co. No Fm. Refs. 78a.
43. Paisley, Lake Co. No Fm. Refs. 78a, 125.
44. Palmer Creek (on Yamhill River S of Dayton), Yamhill Co.
    No. Fm. Refs. 78a.
45. Portland (West Park and City Park), Washington Co. No Fm.
    Refs. 78a, 125.
46. Prineville, Crook Co. No Fm. Refs. 78a, 125.
47. Rainier, Columbia Co. No Fm. Refs. 78a, 125.
48. Rye valley, Baker Co. No Fm. Refs. 78a.
49. Silver Lake, Lake Co. No Fm. Refs. 41m, 78a, 125, 188.
50. St. Paul (mouth of Willamette River), Marion Co.
    No Fm. Refs. 78a.
51, Summer Lake, Lake Co. No Fm. Refs. 78a.
52. Talent (on Anderson Creek), Jackson Co. No Fm. Refs. 78a.
53. Ten Mile Creek (S of Cape Perpetua), Lane Co. No Fm.
    Refs. 78a, 125.
54. The Dalles (on 15 Mile Creek and 10 Mile Creek), Wasco Co.
    No Fm. Refs. 38a, 78a, 125.
55. Willamette valley (general locality). No Fm. Refs. 78a,
    125, 205.
56. Wilsonville, Clackamas Co. No Fm. Refs. 125.
57. Yamhill River (near Dayton), Yamhill Co. No Fm.
    Refs. 38a, 78a, 125.

BIBLIOGRAPHY

1. Addicott, Warren O., 1964. A late Pleistocene invertebrate fauna from southwestern Oregon. Jour. Paleo., v.38, no.4, p.650-661.

a--1972. Neogene molluscan paleontology along the West Coast of North America, 1840-1969. Trends and status. Jour. Paleo., v.46, no.5, p.627-636.

b--1976. Neogene molluscan stages of Oregon and Washington. Amer. Assoc. Petroleum Geol., Bull., v.60, no.12, p.2174.

2. Adegoke, O.S., 1967. A probable pogonophoran from the early Oligocene of Oregon. Jour. Paleo., v.41, no.5, p.1090-1094.

3. Allen, J.E., and Baldwin, E.M., 1944. Geology and coal resources of the Coos Bay quadrangle, Oregon. Oregon Dept. Geol. and Min. Indus., Bull.27, 153p.

4. Allison, Ira S., 1966. Fossil Lake, Oregon - its geology and fossil faunas. Oregon State Univ. Monogr., no.9, 48p.

5. Anderson, Frank M., 1958. Upper Cretaceous of the Pacific Coast. Geol. Soc. America, Memoir no.71, 378p.

6. Applegate, S.P., 1968. A large fossil sand shark of the genus *Odontaspis* from Oregon. Ore Bin, v.30, no.2, p.32-36.

7. Armentrout, J.M., 1967. The Tarheel and Empire Formations; geology and paleontology of the type sections , Coos Bay, Oregon. Univ. of Oregon, Masters, 155p.

8. Arnold, C.A., 1937. Observations on the fossil flora of eastern and southeastern Oregon. Univ. Mich. Mus. Paleont., Contr., pt.1, v.5, no.8, p.79-102.

a--1945. Silicified plant remains from the Mesozoic and Tertiary of western North America. Mich. Acad. Sci., Papers, v.30, p.3-34.

b--1952. Fossil Osmundaceae from the Eocene of Oregon. Palaeontographica B, v.91, p.63-78.

c--1953. Fossil plants of early Pennsylvanian type from central Oregon. Palaeontographica B, v.93, p.61-68.

d--1964. Mesozoic and Tertiary fern evolution evolution and distribution. Torrey Bot. Club, Mem., v.21, no.5, p.58-66.

9. Axelrod, D.I., 1944. The Alvord Creek flora. Carnegie Inst. Wash., Publ.553, p.225-306.

a--1966a. A method for determining the altitudes of Tertiary floras. Paleobotanist, v.14, no.1-3, p.144-171.

b--1966b. Potassium-argon ages of some western Tertiary floras. Amer. Jour. Sci., v.264, no.7, p.497-506.

10. Axelrod, D.I., and Bailey, H., 1969. Paleotemperature analysis of the Tertiary plants. Paleo., paleo., paleo., v.6, p.163-195.

11. Bailey, I.W., and Sinnott, E.W., 1916. The climatic distribution of certain types of angiosperm leaves. Amer. Jour. Bot., v.3, p.24-39.

12. Baldwin, E.M., 1944. Geology and coal resources of the Coos Bay quadrangle, Oregon. Oregon Dept. Geol. and Min. Indus., Bull.27, 159p.

a--1945. Some revisions of the late Cenozoic stratigraphy of the southern Oregon coast. Jour. Geol., v.53, p.35-46.

257

b--1950. Pleistocene history of the Newport, Oregon, region. Geol. Soc. Oregon Country Newsletter, v.18, p.29-30.

c--1964. Geology of the Dallas and Valsetz quadrangles, Oregon. Oregon Dept. Geol. and Min. Indus., Bull.35, 52p.

d--1973. Geology and mineral resources of Coos County, Oregon. Oregon Dept. Geol. and Min. Indus., Bull.80, 82p.

e--1974. Eocene stratigraphy of southwestern Oregon. Oregon Dept. Geol. and Min. Indus., Bull.83, 40p.

f--1976. Geology of Oregon. Rev. ed., Kendall/Hunt Pub. Co., Dubuque, Iowa, 147p.

13. Baldwin, E.M., Brown, R.D., Gair, J.E., and Pease, M.H., 1955. U.S. Geol. Surv. Oil and Gas Inv. Map, OM 155.

14. Barnes, L.G., and Mitchell, E.D., 1975. Late Cenozoic northeast Pacific Phocidae. Conseil Permanent Int. pour l'Explor. de la Mer, Rapports et Proces-verbaux, v.169, p.34-42.

15. Beaulieu, J.D., 1971. Geologic formations of western Oregon. Oregon Dept. Geol. and Min. Indus., Bull.70, 72p.

a--1972. Geologic formations of eastern Oregon. Oregon Dept. Geol. and Min. Indus., Bull.73, 80p.

16. Beeson, J., 1955. The geology of the southern half of the Huntington quadrangle, Oregon. Univ. of Oreg., Masters, 79p.

17. Berggren, W.A., and VanCouvering, J.A., 1974. The late Neogene. Paleo., paleo., paleo., v.16, p.1-218.

18. Berman, D.S., 1976. A new amphisbaenian (Reptilia: Amphisbaenia) from the Oligocene-Miocene John Day Formation, Oregon. Jour. Paleo., v.50, no.1, p.165-174.

19. Black, C.C., 1963. A review of the North American Tertiary Sciuridae. Harvard Univ., Mus. Comp. Zool., Bull.130, p.113-248.

20. Blake, D.B., and Allison, R.C., 1970. A new West American Eocene species of the recent Australian ophiuroid *Ophiocrossota*. Jour. Paleo., v.44, no.5, p.925-927.

21. Boggs, S., Orr, W.N., and Baldwin, E.M., 1973. Petrographic and paleontologic characteristics of the Rickreall limestone (Eocene) of northwestern Oregon. Jour. Sed. Pet., v.43, p.644-654.

22. Bones, T.J., 1979. Atlas of fossil fruits and seeds from north central Oregon. Oregon Mus. Sci. and Indus., Occasional Papers in Natl. Sci., no.1, 6p.

23. Bowman, F.J., 1940. The geology of the north half of Hampton quadrangle, Oregon. Oregon State Univ., Masters, 71p.

24. Brattstrom, B.H., and Sturn, A., 1959. A new species of fossil turtle from the Pliocene of Oregon with notes on other fossil *Clemmys* from western North America. Southern Calif. Acad. Sci., Bull.58, p.65-71.

25. Brodkorb, P., 1958. Birds from the middle Pliocene of McKay, Oregon. Condor, v.60, p.252-255.

a--1961. Birds from the Pliocene of Juntura, Oregon. Florida Scientist, v.24, no.3, p.169-184.

26. Brown, C.E., and Thayer, T.P., 1966. Changes in stratigraphic nomenclature. U.S. Geol. Surv. Bull.1244-A, p.25-29.

27. Brown, R.W., 1937. Additions to some fossil floras of the Western United States. U.S. Geol. Surv. Prof. Paper 186-J, p.163-206.

a--1940. New species and changes of name in some American fossil
     floras. Jour. Wash. Acad. Sci., v.30, p.344-356.
b--1959. A bat and some plants from the upper Oligocene of
     Oregon. Jour. Paleo., v.33, no.1, p.135-139.
c--1975. *Equisetum clarnoi*, a new species based on petrifactions
     from the Eocene of Oregon. Amer. Jour. Bot., v.62, p.410-415.
28. Buffetaut, E., 1979. Jurassic marine crocodilians (Mesosuchia:
     Teleosauridae) from central Oregon: first record in North
     America. Jour. Paleo., v.54, no.1, p.211-215.
29. Cavender, T.M., 1968. Freshwater fish remains from the Clarno
     Formation, Ochoco Mountains of northcentral Oregon. Ore
     Bin, v.30, no.7, p.135-141.
a--1969. An Oligocene mudminnow (Family Umbridae) from Oregon
     with remarks on relationships within the Esocoidei. Mich.
     Univ., Mus. Zool., Occasional Paper 660, 33p.
30. Cavender, T.M., Lundberg, J.G., and Wilson, R.L., 1970. Two
     new fossil refords of the genus *Esox* (Teleostei, Salmoni-
     formes) in North America. Northwest Sci., v.44, no.3,
     p.176, 183.
31. Cavender, T.M., and Miller, R.R., 1972. *Smilodonichthys
     rastrosus*, a new Pliocene salmonid fish from Western United
     States. Univ. of Oregon, Mus. Natl. Hist. Bull.18, 44p.
32. Chaney, R.W., 1918. Ecological significance of the Eagle
     Creek flora of the Columbia River Gorge. Jour. Geol.,
     v.26, 0.577-592.
a--1920. Flora of the Eagle Creek Formation, Washington and
     Oregon. Univ. Chicago, Contrib. from the Walker Mus.,
     v.2, no.5, p.115-151.
b--1925a. A comparative study of the Bridge Creek flora and the
     modern redwood forest. Carnegie Inst. Wash. Publ.349, p.1-22.
c--1925b. The Mascall flora; its distribution and climatic relation.
     Carnegie Inst. Wash., Contrib. to Paleo. no.349, p.25-48.
d--1927. Geology and paleontology of the Crooked River Basin with
     special reference to the Bridge Creek flora. Carnegie Inst.
     Wash., Publ.346, pt.4, p.45-138.
e--1938. The Deschutes flora of eastern Oregon. Carnegie Inst.
     Wash., Contrib. to Paleo., v.476, p.185-216.
f--1944a. The Dalles flora. Carnegie Inst. Wash., Publ.553,
     p.285-321.
g--1944b. The Troutdale flora. Carnegie Inst. Wash., Publ.553,
     p.323-351.
h--1948. Pliocene flora from the Rattlesnake Formation of Oregon.
     Geol. Soc. Amer., Bull. (abstr.), v.159, pt.2, p.1367.
i--1956. The ancient forests of Oregon. Condon Lectures, Eugene,
     Oregon. 56p. (Also in: Carnegie Inst. Wash., Publ.501,
     p.631-648, 1938).
33. Chaney, R.W., and Axelrod, D.I., 1959. Miocene floras of
     the Columbia Plateau. Carnegie Inst. Wash., Publ.617, 237p.
34. Chaney, R.W., and Sanborn, E.I., 1933. The Goshen flora of
     west central Oregon. Carnegie Inst. Wash., Publ.439, 103p.
35. Clifton, H.E., and Boggs, S., 1970. Concave-up pelecypod
     (*Psephidia*) shells in shallow marine sand, Elk River Beds,
     southwestern Oregon. Jour. Sed. Pet., v.40, no.3, p.888-897.

36. Cockerell, T.D.A., 1927. Tertiary fossil insects from eastern Oregon. Carnegie Inst., Wash., Contrib. to Paleo., no.346, p.64-65.

37. Colbert, E.H., 1938. Pliocene peccaries from the Pacific Coast region of North America. Carnegie Inst., Wash., Publ.487, p.243-269.

38. Condon, T., 1906. A new fossil pinniped (*Desmatophoca oregonensis*) from the Miocene of the Oregon coast. Univ. of Oregon, Bull. suppl., v.3, p.1-14.

a--1902. The two Islands. J.K. Gill Co., Portland, Oregon, 211p.

39. Conrad, T.A., 1848. Fossil shells from Tertiary deposits on the Columbia River, near Astoria. Amer. Jour. Sci., 2d ser., v.5, p.432-433.

40. Cooper, G.A., 1957. Permian brachiopods from central Oregon. Smithsonian Misc. Coll., v.134, no.12, 79p.

41. Cope, E.D., 1878. Descriptions of new vertebrata from the upper Tertiary formations of the West. Amer. Phil. Soc., Proc.17, p.219-231.

a--1879. Observations on the faunae of the Miocene Tertiaries of Oregon. U.S. Geog. and Geol. Surv., Bull.5, p.55-59.

b--1880. Extinct cats of North America. Amer. Nat., v.14, p.833-858.

c--1882. On the Nimravidae and Canidae of the Miocene period. U.S. Geog. and Geol. Surv., Bull.6, p.165-181.

d--1883a. Extinct dogs of North America. Amer. Nat., v.17, p.235-249.

e--1883b. Extinct Rodentia of North America. Amer. Nat., v.17, p.43-57; p.165-174; p.370-381.

f--1883c. A new Pliocene formation in the Snake River Valley. Amer. Nat., v.17, p.867-868.

g--1883d. On the fishes of the Recent and Pliocene lakes of the western part of the Great Basin and of the Idaho Pliocene lake. Acad. Nat. Sci., Philadelphia, Proc., p.134-166.

h--1883e. Vertebrata of the Tertiary formations of the West. U.S. Geog. and Geol. Surv., Report, v.3, p.762-1002.

i--1886. Phylogeny of the Camelidae. Amer. Nat., v.20, p.611-624.

j--1888. On the Dictyolinae of the John Day Miocene of North America. Amer. Phil. Soc., Proc., v.25, p.62-79.

k--1889a. The Edentata of North America. Amer. Nat., v.23, p.657-665.

l--1889b. On a species of *Plioplarchus* from Oregon. Amer. Nat., v.12, p.625-626.

m--1889c. The Silver Lake of Oregon and its region. Amer. Nat., v.23, p.970-982.

42. Dake, H.C., 1969. Oregon Tempskya locality. Gems and minerals, no.384, p.62-63.

43. Dall., W.H., 1909. Contributions to the Tertiary paleontology of the Pacific Coast. U.S.-Geol. Surv., Prof. Paper 59, 216p.

44. Daugherty, L.F., 1951. The molluscs and foraminifera of Depoe Bay, Oregon. Univ. of Oregon, Masters, 77p.

45. David, L.R., 1956. Tertiary anacanthin fishes from California and the Pacific Northwest; their paleoecological significance. Jour. Paleo., v.30, no.3, p.568-607.

46. Dawson, M.R., 1965. Oreolagus and other lagomorpha (Mammalia) from the Miocene of Colorado, Wyoming and Oregon. Colorado Univ. Studies in Earth Sci., no.1, 36p.

47. Detling, L.E., 1968. Historical background of the flora of the Pacific Northwest. Univ. of Oregon, Mus. Nat. Hist., Bull.13, 57p.

48. Dickinson, W.R., and Vigrass, L.W., 1965. Geology of the Suplee-Izee area, Crook, Grant, and Harney Counties, Oregon. Oregon Dept. Geol. and Min. Indus., Bull.58, 108p.

49. Diller, J.S., 1896. A geological reconnaissance in northwestern Oregon. U.S. Geol. Surv., 17th Ann. Rept., pt.1, p.441-520.

a--1898. Description of the Roseburg quadrangle. U.S. Geol. Surv., Geol. Atlas, Roseburg Folio no.49.

b--1899. The Coos Bay coalfield, Oregon. U.S. Geol. Surv. 19th Ann. Rept., pt.3, p.309-370.

c--1901. Description of the Coos Bay quadrangle, Oregon. U.S. Geol. Surv., Geol. Atlas, Folio 73.

d--1902. Topographic development of the Klamath Mountains. U.S. Geol. Surv., Bull.196, 69p.

e--1907. The Mesozoic sediments of southwestern Oregon. Amer. Jour. Sci., 4th ser., v.23, p.401-421.

f--1908. Strata containing the Jurassic flora of Oregon. Geol. Soc. Amer., Bull.19, p.367-402.

50. Dodds, B.R., 1963. The relocation of geologic locales in Oregon. Ore Bin, v.25, no.7, p.113-128.

51. Dott, R.H., 1966a. Eocene deltaic sedimentation at Coos Bay, Oregon. Jour. Geol., v.74, no.4, p.373-420.

b--1966b. Late Jurassic unconformity exposed in southwestern Oregon. Ore Bin, v.28, no.5, p.85-97.

52. Downs, T., 1952. A new mastodont from the Miocene of Oregon. Univ. Calif., Publ. Geol. Sci., v.29, no.1, p.1-20.

a--1956. The Mascall fauna from the Miocene of Oregon. Univ. Calif., Publ. Geol. Sci., v.31, p.199-354.

53. Durham, J.W., Harper, H., and Wilder, B., 1942. Lower Miocene in the Willamette Valley, Oregon. Geol. Soc. Amer., Bull.53, pt.2, p.1817.

54. Eaton, G.T., 1922. The John Day Felidae in the Marsh Collection. Amer. Jour. Sci., ser.5, v.4, p.425-452.

55. Emlong, D.R., 1966. A new archaic cetacean from the Oligocene of northwest Oregon. Univ. of Oregon. Mus. Nat. Hist., Bull.3, 51p.

56. Eubanks, W., 1960. Fossil woods of the Thomas Creek area, Linn County, Oregon. Ore Bin, v.22, no.7, p.65-69.

a--1962. The fossil flora of Thomas Creek. Ore Bin, v.24, no.2, p.26-27.

57. Evernden, J.F., and James, G.T., 1964. Potassium-argon dates of the Tertiary floras of North America. Amer. Jour. Sci., v.262, no.8, p.945-974.

58. Evernden, J.F., et al., 1964. Potassium-argon dates and the Cenozoic mammalian chronology of North America. Amer. Jour. Sci., v.262, p.145-198.

59. Feldmann, R.M., 1974. *Haploparia riddlensis*, a new species of lobster (Decapoda: Nephropidae) from the Days Creek Formation (Hauterivian, lower Cretaceous) of Oregon. Jour. Paleo., v.48, no.3, p.586-593.

60. Fisher, R.V., and Rensberger, J.M., 1972. Physical stratig-
    raphy of the John Day Formation, central Oregon. Univ.
    Calif., Publ. Geol. Sci., v.101, p.1-45.
61. Fontaine, W.M., 1905. Notes on some fossil plants from the
    Shasta Group of California and Oregon. U.S. Geol. Surv.,
    Monograph 48, pt.1, p.221-273.
62. Fouch, T.D., 1968. Geology of the northwest quarter of the
    Brogan quadrangle, Malheur County, Oregon. Univ. of Oregon,
    Masters, 62p.
63. Frey, R.W., and Cowles, J.G., 1972. The trace fossil *Tisoa*
    in Washington and Oregon. Ore Bin, v.34, no.7, p.113-119.
64. Fry, W.E., 1973. Giant fossil tortoise of genus *Geochelone*
    from the late Miocene-early Pliocene of north central
    Oregon. Northwest Sci., v.17, no.4, p.239-249.
65. Furlong, E.L., 1932. Distribution and description of skull
    remains of the Pliocene antelope *Sphenophalos* from the
    northern Great Basin province. Carnegie Inst., Wash.,
    Publ. 418, p.27-36.
66. Gazin, C.L., 1932. A Miocene mammalian fauna from south-
    eastern Oregon. Carnegie Inst., Wash., Publ.418, p.39-86.
67. Gilluly, J., 1937. Geology and mineral resources of the
    Baker quadrangle, Oregon. U.S. Geol. Surv. Bull.879, 119p.
68. Graham, A.K., 1963. Systematic revision of the Sucker Creek
    and Trout Creek Miocene floras of southeastern Oregon.
    Amer. Jour. Botany, v.50, no.9, p.921-936.
a--1965. The Sucker Creek and Trout Creek Miocene floras of
    southeastern Oregon. Kent State Univ. Research Series 9
    (Bull.53, no.12), 147p.
69. Gregory, I., 1968. The fossil woods near Holley in the
    Sweet Home petrified forest, Linn County, Oregon. Ore
    Bin, v.30, no.4, p.57-76.
a--1969. Fossilized palm wood in Oregon. Ore Bin, v.31, no.5,
    p.93-110.
70. Hall, E.R., 1944. A new genus of American Pliocene badger,
    with remarks on the relationships of badgers of the Northern
    Hemisphere. Carnegie Inst., Wash., Contrib. to Paleo.,
    Publ.551, p.9-23.
71. Hancock, L., 1962. "The new mammal beds." Geol. Soc. Oregon
    Country, Newsletter, v.28, no.6, p.39-40.
72. Hanna, G.D., 1920. Fossil mollusks from the John Day Basin
    in Oregon. Univ. of Oregon, Publ., v.1, no.6, 8p.
a--1922. Fossil freshwater mollusks from Oregon contained in the
    Condon Museum of the University of Oregon. Univ. of
    Oregon, Publ., v.1, no.12, 22p.
b--1963. Some Pleistocene and Pliocene freshwater mollusca
    from California and Oregon. Calif. Acad. Sci., Occasional
    Paper no.43, 20p.
73. Hannibal, H., 1922. Notes on Tertiary sirenians of the
    genus *Desmostylus*. Jour. Mammal., v.3, p.238-240.
74. Hansen, H.P., 1942. The influence of volcanic eruptions
    upon post-Pleistocene forest succession in central Oregon.
    Amer. Jour. Botany, no.29, p.214-219.
a--1946. Post glacial forest succession and climate in the
    Cascades. Amer. Jour. Sci., v.244, p.710-734.

b--1947a. Post glacial forest succession, climate, and chronology in the Pacific Northwest. Amer. Philos. Soc., Trans., v.37, pt.1, 130p.

c--1947c. Post glacial vegetation of the northern Great Basin. Amer. Jour. Botany, v.34, p.164-171.

75. Hansen, H.P., and Packard, E.L., 1949. Pollen analysis and the age of proboscidian bones near Silverton, Oregon. Ecology, v.30, no.4, p.461-468.

76. Harper, H.E., 1946. Geology of the Molalla quadrangle. Oregon State Univ., Masters, 29p.

77. Hauck, S.M., 1962. Geology of the southwest quarter of the Brownsville quadrangle, Oregon. Oregon Univ., Masters, 82p.

78. Hay, O.P., 1903. Two new species of fossil turtles from Oregon. Univ. Calif., Publ. Geol. Sci., v.3, no.10, p.237-241.

a--1908. The fossil turtles of North America. Carnegie Inst., Wash., Publ.75, p.1-568.

b--1916. A contribution to the knowledge of the extinct sirenian *Desmostylus hesperus*, Marsh. U.S. Natl. Mus. Proc. v.49, p.381-397.

c--1927. The Pleistocene of the western region of North America and its vertebrated animals. Carnegie Inst., Wash., Publ.322B, 346p.

79. Hergert, H.L., 1961. Plant fossils in the Clarno Formation, Oregon. Ore Bin, v.23, no.6, p.55-62.

80. Hickman, C.J.S., 1969. The Oligocene marine molluscan fauna of the Eugene Formation in Oregon. Univ. of Oregon, Mus. Nat. Hist. Bull., no.16, 112p.

a--1972. Review of the bathyal gastropod genus *Phanerolepida* (Homalopomatinae) and description of a new species from the Oregon Oligocene. Veliger, v.15, no.2, p.89-112.

b--1974. *Nehalemia hieroglyphica*, a new genus and species of Archaeogastropod (Turbinidae: Homalopomatinae) from the Eocene of Oregon. The Veliger, v.17, no.2, p.89-91.

c--1976. *Pleurotomaria* (Archaeogastropoda) in the Eocene of the north-eastern Pacific; a review of the Cenozoic biogeography and ecology of the genus. Jour. Paleo., v.50, no.6, p.1090-1102.

81. Hodge, E.T., 1932. New evidence of the age of the John Day Formation. Geol. Soc. Amer., Bull., v.43, p.695-702.

82. Hogenson, G.M., 1964. Geology and ground water of the Umatilla River Basin, Oregon. U.S. Geol. Surv., Water Supply Paper 1620, 162.

83. Hoover, L., 1963. Geology of the Anlauf and Drain quadrangles, Douglas and Lane Counties, Oregon. U.S. Geol. Surv., Bull.1122-D, 62p.

84. Hopkins, W.S., 1967. Palynology and its paleoecological applications in the Coos Bay area, Oregon. Ore Bin, v.29 no.9, p.161-183.

85. Howard, H., 1946. A review of the Pleistocene birds of Fossil Lake, Oregon. Carnegie Inst. Wash., Publ.551, p.141-195.

a--1964. A new species of the "Pigmy Goose", *Anabernicula*, from the Oregon Pleistocene, with a discussion of the genus. Amer. Mus. Nat. Hist., Novitates, no.2200, 14p.

86. Howe, H.V., 1922. Faunal and stratigraphic relationships of the Empire Formation, Coos Bay, Oregon. Calif. Univ., Dept. Geol. Sci., Bull., v.14, p.85-114.

87. Hoxie, L.R., 1965. The Sparta Flora from Baker County, Oregon. Northwest Sci., v.39, no.1, p.26-35.
88. Hutchinson, J.H., 1966. Notes on some upper Miocene shrews from Oregon. Univ. of Oregon, Mus. Nat. Hist., Bull.2, 23p.
a--1968. Fossil Talpidae (Insectivora, Mammalia) from the later Tertiary of Oregon. Univ. of Oregon, Mus. Nat. Hist., Bull.11, 117p.
89. Imlay, R.W., 1967. The Mesozoic pelecypods *Otapira* and *Lupherella*, Imlay, new genus in the United States. U.S. Geol. Surv., Prof. Paper 573-B, p.1-11.
a--1968. Lower Jurassic (Pliesbachian and Toarcian) ammonites from eastern Oregon and California. U.S. Geol. Surv., Prof. Paper 593-C, p.1-51.
b--1973. Middle Jurassic (Bajocian) ammonites from eastern Oregon. U.S. Geol. Survey, Prof. Paper 756, 100p.
90. Imlay, R.W., and Jones, D.L., 1970. Ammonites from the *Buchia* zones in northwestern California and southwestern Oregon. U.S. Geol. Surv., Prof. Paper 647-B, p.1-59.
91. Imlay, R.W., Dole, H.M., Wells, F.G., and Peck, D.L., 1959. Relations of certain upper Jurassic and lower Cretaceous formations in southwestern Oregon. Amer. Assoc. Petroleum Geol., Bull., v.43, p.2770-2785.
92. Jehl, J.R., 1967. Pleistocene birds from Fossil Lake, Oregon. Condor, v.69, no.1, p.24-27.
93. Johnson, A.M., 1931. Taxonomy of the flowering plants. The Century Co., New York, 864p.
94. Johnson, J.G., and Klapper, G., 1978. Devonian brachiopods and conodonts from central Oregon. Jour. Paleo., v.52, no.2, p.295-299.
95. Jones, D.L., 1960. Lower Cretaceous (Albian) fossils from southwestern Oregon and their paleogeographic significance. Jour. Paleo., v.34, no.1, p.152-160.
a--1969. *Buchia* zonation in the Myrtle Group, southwestern Oregon. Geol. Soc. Amer., Cordilleran Sect., Abs. with Programs, pt.3, p.31-32.
96. Jordan, D.S., 1907. The fossil fishes of California; with supplementary notes on other species of extinct fishes. Calif. Univ., Dept. Geol. Sci., Bull.5, p.95-144.
97. Jordan, D.S., and Hannibal, H., 1923. Fossil sharks and rays of the Pacific slope of North America. Southern Calif. Acad. Sci., Bull., v.22, p.27-68.
98. Keen, A.M., and Coan, E., 1974. Marine molluscan genera of Western North America; an illustrated key. 2d ed., Stanford Univ. Press, Stanford, Calif., 208p.
99. Kimmel, P.G., 1975. Fishes of the Miocene-Pliocene Deer Butte Formation, southeast Oregon. Univ. Mich., Mus. Paleo., Papers on Paleo., no.14, p.69-87.
100. Kittleman, L.R., 1965. Cenozoic stratigraphy of the Owyhee region, southeastern Oregon. Univ. of Oregon, Mus. Nat. Hist., Bull.1, 45p.
101. Kleweno, W.P., and Jeffords, R.M., 1961. Devonian rocks in the Suplee area of central Oregon. Ore Bin, v.23, no.5, p.50.

102. Klucking, E.P., 1964. An Oligocene flora from the western Cascades. Univ. Calif., PhD., 372p.

103. Knowlton, F.H., 1900. Fossil plants associated with the lavas of the Cascade Range. U.S. Geol. Surv. 20th Ann. Rept., pt.3, p.37-64.

a--1902. Fossil flora of the John Day Basin, Oregon. U.S. Geol. Surv., Bull.204, 127p.

b--1910. Jurassic age of the "Jurassic flora of Oregon." Amer. Jour. Sci., 4th ser., v.30, p.33-64.

104. Koch, J.G., 1966. Late Mesozoic stratigraphy and tectonic history, Port Orford-Gold Beach area, southwestern Oregon coast. Amer. Assoc. Petroleum Geol., Bull., v.50, no.1, p.25-71.

105. Koch, J.G., and Camp, C.L., 1966. Late Jurassic ichthyosaur from Sisters Rocks coastal southwestern Oregon. Ore Bin, v.28, no.3, p.65-68.

106. Kooser, M.A., and Orr, W.N., 1973. Two new decapod species from Oregon. Jour. Paleo., v.47, no.6, p.1044-1046.

107. Krasilov, V.A., 1975. Paleoecology of terrestrial plants. New York, Wiley Pub.

108. Lakhanpal, R.N., 1958. The Rujada flora of west central Oregon. Univ. Calif., Publ. Geol. Sci., v.35, p.1-66.

109. Leffler, S.R., 1964. Fossil mammals from the Elk River Formation, Cape Blanco, Oregon. Jour. Mammal., v.45, p.53-61.

110. Leidy, J., 1870. Remarks on a collection of fossils from ... Thomas Condon. Acad. Nat. Sci., Philadelphia, Proc., v.22, p.111-113.

a--1871. Remarks on fossils from Oregon. Acad. Nat. Sci., Philadelphia, Proc., v.24, p.247-248.

b--1873. Contributions to the extinct vertebrate fauna of the western territories. U.S. Geol. Surv., Rept. of Territories (Hayden), v.1, 358p.

111. Lesquereux, L., 1888. Recent determinations of fossil plants from Kentucky, Louisiana and Oregon. U.S. Natl. Mus., Proc., v.11, p.11-38.

112. Lewis, R.Q., 1950. The geology of the southern Coburg Hills including the Springfield-Goshen area. Univ. of Oregon, Masters, 58p.

113. Lowry, W.D., 1940. Geology of the Bear Creek area, Crook and Deschutes Counties, Oregon. Masters, Oregon State Univ., 78p.

a--1943. The geology of the northeast quarter of the Ironside Mountain quadrangle, Baker and Malheur Counties, Oregon. Univ. Rochester, PhD.

114. Lowther, J.S., 1967. A new Cretaceous flora from southwestern Oregon (abstr.). Northwest Sci., v.41, no.1, p.54.

115. Lull, R.S., 1921. New camels in the Marsh Collections. Amer. Jour. Sci., ser.5, v.1, no.5, p.392-404.

116. Lundberg, J.G., 1975. The fossil catfishes of North America. Univ. Mich., Mus. Paleo., Papers on Paleo., no.11, p.1-51.

117. Lupher, R.I., 1941. Jurassic stratigraphy of central Oregon. Geol. Soc. Amer., Bull., v.52, pt.1, p.219-269.

118. Lupher, R.I., and Packard, E.L., 1930. The Jurassic and Cretaceous rudistids of Oregon. Univ. of Oregon, Publ. Geol., v.1, no.3, p.203-212.

119. Mamay, S.H., and Read, C.B., 1956. Additions to the flora of the Spotted Ridge Formation in central Oregon. U.S. Geol. Surv., Prof. Paper 274-I, p.211-226.

120. Marsh, O.C., 1873. Notice of new Tertiary mammals. Amer. Jour. Sci., ser.3, v.5, p.407-410; p.485-488.

a--1874. Notice of new equine mammals from the Tertiary formations. Amer. Jour. Sci., ser.3, v.7, p.147-158.

b--1894. Eastern division of the Miohippus beds, with notes on some of the characteristic fossils. Amer. Jour. Sci., ser.3, v.48, p.91-94.

c--1895. Reptilia of the Baptanodon beds. Amer. Jour. Sci., v.50, no.3, p.405-406.

121. Mason, H.L., 1927. Fossil records of some west American conifers. Carnegie Inst., Wash., Contrib. to Paleo., no.346, p.139-158.

122. Matthew, W.D., 1899. A provisional classification of the freshwater Tertiary of the West. Amer. Mus. Nat. Hist., Bull.12, p.19-75.

a--1909. Faunal lists of the Tertiary mammalia of the West. U.S. Geol. Surv., Bull.361, p.91-138.

123. Mawby, J.E., 1960. A new American occurrence of *Heterosorex* Gaillard. Jour. Paleo., v.34, p.950-956.

124. Maxson, J.H., 1928. *Merychippus isonesus* (Cope) from the later Tertiary of the Crooked River Basin, Oregon. Carnegie Inst., Wash., Contrib. to Paleo., no.393, p.55-58.

125. McCornack, E.C., 1914. A study of Oregon Pleistocene. Univ. of Oregon, Bull., n.s., v.12, no.2, 16p.

126. McGintie, H.D., 1933. The Trout Creek flora of southeastern Oregon. Carnegie Inst., Wash., Publ.416, p.21-68.

127. McKee, T.M., ?1970. Preliminary report on the fossil fruits and seeds from in the mammal quarry of the lower Tertiary Clarno Formation, Oregon. Oregon Mus. Sci. and Indus., Student Res. Center, unpubl. rept., 17p.

128. Mellett, James S., 1969. A skull of *Hemipsalodon* (Mammalia, order Deltatheridia) from the Clarno Formation of Oregon. Amer. Mus. Novitates, no.2387, 19p.

129. Mendenhall, W.C., 1909. A coal prospect on Willow Creek, Morrow County, Oregon. U.S. Geol. Surv., Bull. 341-C, p.406-408.

130. Merriam, C.W., 1941. Fossil Turritellas from the Pacific Coast region of North America. Univ. Calif., Publ. Geol. Sci., v.26, no.1, p.1-214.

a--1942. Carboniferous and Permian corals from central Oregon. Jour. Paleo., v.16, p.372-381.

131. Merriam, C.W., and Berthiaume, S.A., 1943. Late Paleozoic formations of central Oregon. Geol. Soc. Amer., Bull., v.54, pt.1, p.145-171.

132. Merriam, J.C., 1901. Contribution to the geology of the John Day Basin. Univ. Calif., Publ. Geol. Sci., v.9, p.269-314.

a--1906. Carnivora from the Tertiary formations of the John Day region. Calif. Univ., Publ. Geol. Sci., v.5, p.1-64.

b--1913. Tapir remains from the late Cenozoic beds of the Pacific Coast region. Calif. Univ., Publ. Geol. Sci., v.7, no.9, p.169-175.

c--1916. Mammalian remains from a late Tertiary formation at Ironside, Oregon. Calif. Univ., Publ. Geol. Sci., v.10, no.9, p.129-135.

d--1917. Relationships of Pliocene mammalian faunas from the Pacific Coast and Great Basin provinces of North America. Calif. Univ., Publ. Geol. Sci., v.10, no.22, p.421-443.

e--1930. *Allocyon*, a new canid genus from the John Day beds of Oregon. Calif. Univ., Publ. Geol. Sci., v.19, no.9, p.229-244.

133. Merriam, J.C., and Sinclair, W.J., 1907. Tertiary faunas of the John Day region. Calif. Univ., Publ. Geol. Sci., v.5, no.11, p.171-205.

134. Merriam, J.C., and Stock, C., 1927. A hyaenarctid bear from the later Tertiary of the John Day Basin, Oregon. Carnegie Inst., Wash., Publ.346, p.39-44.

135. Merriam, J.C., Stock, C., and Moody, C.L., 1925. The Pliocene Rattlesnake Formation and fauna of eastern Oregon, with notes on the geology of the Rattlesnake and Mascall deposits. Carnegie Inst., Wash., Publ.347, p.43-92.

136. Merriam, J.C., and Gilmore, C.W., 1928. An ichthyosaurian reptile from marine Cretaceous of Oregon. Carnegie Inst., Wash., Contrib. to Paleo., no.393, p.1-4.

137. Meyer, H., 1973. The Oligocene Lyons flora of north-western Oregon. Ore Bin, v.35, no.3, p.37-51.

138. Miller, A.H., 1911. Additions to the avifauna of the Pleistocene deposits at Fossil Lake, Oregon. Calif. Univ., Publ. Geol. Sci., v.6, no.4, p.79-97.

a--1912. Contributions to avian paleontology from the Pacific Coast of North America. Calif. Univ., Publ. Geol. Sci., v.7, no.5, p.61-115.

b--1931. An auklet from the Eocene of Oregon. Calif. Univ., Publ. Geol. Sci., v.20, p.23-26.

c--1944. Some Pliocene birds from Oregon and Idaho. Condor, v.46, p.25-32.

d--1957. Bird remains from an Oregon Indian midden. Condor, v.59, p.59-63.

139. Miller, R.R., 1965. Quaternary freshwater fishes of North America. In: Quaternary of the United States. Princeton, N.J., Princeton Univ. Press, p.569-581.

140. Mitchell, E., 1966. Faunal succession of extinct North Pacific marine mammals. Norsk Hvalfangst-Tidende, no.3, p.47-60.

141. Mobley, B.J., 1956. Geology of the southwest quarter of the Bates quadrangle, Oregon. Univ. of Oregon, Masters, 66p.

142. Moore, E.J., 1963. Miocene marine mollusks from the Astoria Formation in Oregon. U.S. Geol. Surv., Prof. Paper 419, 190p.

a--1971. Fossil mollusks of coastal Oregon. Oregon State Univ., Monograph no.10, 64p.

b--1976. Oligocene marine mollusks from the Pittsburg Bluff Formation in Oregon. U.S. Geol. Surv., Prof. Paper 922, p.1-66.
143. Moore, E.J., and Vokes, H.E., 1953. Lower Tertiary crinoids
from northwestern Oregon. U.S. Geol. Surv., Prof. Paper
233-E, p.113-148.
144. Morrison, R.F., 1964. Upper Jurassic mudstone unit named
in Snake River Canyon, Oregon-Idaho boundary. Northwest
Sci., v.38, no.3, p.83-87.
145. Munthe, J., and Coombs, M.C., 1979. Miocene dome-skulled
chalicotheres (Mammalia, Perissodactyla) from the western
United States: a preliminary discussion of a bizarre
structure. Jour. Paleo., v.53, no.1, p.77-91.
146. Nations, J.D., 1970. The family Cancridae and its fossil
record on the West Coast of North America. Univ. Calif., PhD.
147. Newberry, J.S., 1882. Brief descriptions of plants,
chiefly Tertiary, from western North America. U.S. Natl.
Mus., Proc.5, p.502-514.
148. Newcomb, R.C., 1958. Yonna Formation of the Klamath River
Basin, Oregon. Northwest Sci., v.32, no.2, p.41-48.
149. Niem, A.R., VanAtta, R., et al., 1973. Cenozoic stratigraphy,
Oregon-Washington: road log. In: Geologic field trips in
northern Oregon and southern Washington. Trip no.3. Oregon
Dept. Geol. and Min. Indus., Bull., no.77, p.75-132.
150. Nomland, J., 1917. New fossil corals from the Pacific Coast.
Univ. Calif., Publ. Geol. Sci. v.10, no.13, p.185-190.
151. Oliver, E., 1934. A Miocene flora from the Blue Mountains,
Oregon. Carnegie Inst., Wash., Publ.455, pt.1, p.1-27.
152. Orr, W.N., and Faulhaber, J., 1975. A middle Tertiary
cetacean from Oregon. Northwest Sci., v.49, p.174-181.
153. Orr, W.N., and Kooser, M.A., 1971. Oregon Eocene decapod
Crustacea. Ore Bin, v.33, p.119-129.
154. Osborn, H.F., 1902. A remarkable new mammal from Japan:
its relationship to the California genus *Desmostylus*
Marsh. Science, n.s., v.16, p.713-714.
a--1936. Proboscidea; a monograph of the discovery, evolution,
migration and extinction of the mastodonts and elephants
of the world. New York, Amer. Mus. Press, 2v. in 1675p.
155. Packard, E.L., 1921. The Trigoniae from the Pacific Coast of
North America. Oregon Univ. Publ., v.1, no.9, p.1-40.
a--1923. An aberrant oyster from the Oregon Eocene. Oregon
Univ. Publ., v.2, no.4, p.1-6.
b--1940. A new turtle from the marine Miocene of Oregon. Oregon
State Univ., Studies in Geology, no.2, 31p.
c--1947. Fossil baleen from the Pliocene of Cape Blanco, Oregon.
Oregon State Univ., Studies in Geology, v.5, p.1-12.
d--1947a. A fossil sea lion from Cape Blanco, Oregon. Oregon
State Univ., Studies in Geology, v.6, p.13-22.
e--1947b. A pinniped humerus from the Astoria Miocene of Oregon.
Oregon State Univ., Studies in Geology, v.7, p.23-32.
f--1952. Fossil edentates of Oregon. Oregon State Univ.,
Studies in Geology, v.8, 15p.
g--1956. An *Engonoceras*. Jour. Paleo., v.30, no.2, p.398-402.

156. Packard, E.L., and Jones, D.L., 1962. A new species of
     Anisoceras merriami from Oregon. Jour. Paleo., v.36,
     no.5, p.1047-1050.
157. Packard, E.L., and Jones, D.L., 1965. Cretaceous pelecypods
     of the genus *Pinna* from the Pacific Coast region of North
     America. Jour. Paleo., v.39, no.5, p.910-905.
158. Packard, E.L., and Kellogg, A.R., 1934. A new cetothere from
     the Miocene Astoria Formation of Newport, Oregon. Carnegie
     Inst., Wash., Publ.477, p.1-62.
159. Peck, D.L., 1964. Geologic reconnaissance of the Antelope-
     Ashwood area, north central Oregon. U.S. Geol. Surv.,
     Bull. 1161-D, 26p.
160. Peck, D.L., Imlay, R.W., and Popenoe, W.P., 1956. Upper
     Cretaceous rocks of parts of southwestern Oregon and
     northern California. Amer. Assoc. Petroleum Geol.,
     Bull., v.40, no.8, p.1968-1984.
161. Peck, D.L., et al., 1964. Geology of the central and
     northern parts of the Western Cascade Range in Oregon.
     U.S. Geol. Surv., Prof. Paper 449, 56p.
162. Peterson, J.V., 1964. Plant fossils of the Clarno Formation.
     Earth Sci., v.17, no.1, p.11-15.
163. Pigg (West), J.H., 1961. Lower Tertiary sedimentary rocks
     in the Pilot Rock and Heppner areas, Oregon. Univ. of
     Oregon, Masters, 67p.
164. Popenoe, W.P., Imlay, R.W., and Murphy, M.A., 1960.
     Correlation of the Cretaceous formations of the Pacific
     Coast (United States and northwestern Mexico). Geol. Soc.
     Amer., Bull., v.71, p.1491-1540.
165. Quaintance, C.W., 1969. *Mylodon*, furthest north in Pacific
     Northwest. Amer. Midland Naturalist, v.81, no.2, p.593-594.
166. Ramp, L., 1969. Dothan (?) fossils discovered. Ore Bin,
     v.31, no.12, p.245-246.
167. Rathbun, M.J., 1926. The fossil stalk-eyed Crustacea of the
     Pacific slope of North America. U.S. Natl. Mus., Bull.
     138, 155p.
168. Ray, C.E., 1976. Fossil marine mammals of Oregon. Systematic
     Zool., v.25, no.4, p.420-436.
169. Read, C.B., and Brown, R.W., 1937. American Cretaceous
     ferns of the genus *Tempskya*. U.S. Geol. Surv., Prof Paper
     186-F, p.127-128.
170. Read, C.B., and Merriam, C.W., 1940. A Pennsylvanian flora
     from central Oregon. Amer. Jour. Sci., v.238, no.2,
     p.107-111.
171. Rensberger, J.M., 1971. Entophychine pocket gophers
     (Mammalia, Geomyoidea) of the early Miocene John Day
     Formation, Oregon. Calif. Univ., Publ. Geol. Sci.,
     v.90, 209p.
a--1973. Pleurolicine rodents (Geomyoidea) of the John Day
     Formation, Oregon, and their relationships to taxa from
     the early and middle Miocene, South Dakota. Calif.
     Univ., Publ. Geol. Sci., v.102, 95p.
b--1976. John Day Fossil Beds National Monument. Report to
     National Park Service. Unpubl., 61p.

172. Repenning, C.A., 1967. Subfamilies and genera of the Soricidae. U.S. Geol. Surv., Prof. Paper 565, 74p.

a--1968. Mandibular musculature and the origin of the subfamily Arvicolinae (Rodentia). Acta Zool. Cracoviensia, v.13, no.3, p.19-72.

173. Repenning, C.A., and Tedford, R., 1977. Otaroid seals of the Neogene. U.S. Geol. Surv., Prof. Paper 992, p.1-93.

174. Richardson, H.E., 1950. The geology of the Sweet Home petrified forest. Univ. of Oregon, Masters, 44p.

175. Romer, A.S., 1971. Vertebrate paleontology, Univ. of Chicago Press, 661p.

176. Sada, K., and Danner, W.R., 1973. Late lower Carboniferous *Eostaʒella* and *Hexaphyllia* from central Oregon, USA. Paleo. Soc. Japan, Trans. and Proc., no.91, p.151-160.

177. Sanborn, E.I., 1937. The Comstock flora of west central Oregon. Carnegie Inst., Wash., Publ. 465, p.1-28.

a--1947. The Scio flora of western Oregon. Oregon State Univ., Studies in Geology, v.4, p.1-47.

178. Savage, N.M., and Amundson, C.T., 1979. Middle Devonian (Givetian) conodonts from central Oregon. Jour. Paleo., v.53, no.6, p.1395-1400.

179. Scharf, D.W., 1935. A Miocene mammalian fauna from Succor Creek, southeastern Oregon. Carnegie Inst., Wash., Publ.453, p.97-118.

180. Schenck, H.G., 1928. Stratigraphic relations of western Oregon Oligocene formations. Univ. Calif. Publ. Geol. Sci., v.18, 50p.

a--1931. The stratigraphy and paleontology of the Triassic of the Suplee region of central Oregon. Oregon Univ., MS., 53p.

b--1936. Nuculid bivalves of the genus *Acila*. Geol. Soc. Amer., Special Paper 4, 149p.

181. Schlicker, H.G., 1962. The occurrence of Spencer Sandstone in the Yamhill quadrangle, Oregon. Ore Bin, v. 24, no.11, p.173-184.

182. Schultz, C., and Falkenbach, C.H., 1968. The phylogeny of the oreodonts. Amer. Mus. Nat. Hist., Bull., v.139, 498p.

183. Scott, R.A., 1954. Fossil fruits and seeds from the Eocene Clarno Formation of Oregon. Palaeontographica B, v.96, p.66-97.

184. Scott, W.B., 1937. A history of land mammals in the Western Hemisphere. New York, Macmillan.

185. Shimer, H.W., and Shrock, R.R., 1959. Index fossils of North America. 6th ed., N.Y., Wiley, 837p.

186. Shotwell, J.A., 1951. A fossil sea-lion from Fossil Point, Oregon. Oregon Acad. Sci., Proc., v.2, p.97 (abstr.).

a--1956. Hemphillian mammalian assemblage from northeastern Oregon. Geol. Soc. Amer., Bull., v.67, pt.1, p.717-738.

b--1958a. Evolution and biogeography of the aplodontid and mylagaulid rodents. Evolution, v.12, p.451-484.

c--1958b. Inter-community relationships in Hemphillian (Mid-Pliocene) mammals. Ecology, v.39, no.2, p.271-282.

d--1961. Late Tertiary biogeography of horses in the northern
    Great Basin. Jour. Paleo., v.35, no.1, p.203-217.
e--1963. The Juntura Basin: studies in earth history and paleo-
    ecology. Amer. Philos. Soc., Trans., v.53, 77p.
f--1964. Community succession in mammals of the late Tertiary.
    In: Approaches to paleoecology, J. Imbrie and N. Newell,
    ed., Wiley Pub., N.Y., p.135-150.
g--1967a. Late Tertiary geomyoid rodents of Oregon. Univ. of
    Oregon, Mus. Nat. Hist., Bull.9, 51p.
h--1967b. *Peromyscus* of the late Tertiary in Oregon. Univ. of
    Oregon, Mus. Nat. Hist., Bull.5, 35p.
i--1968a. Miocene mammals of southeastern Oregon. Univ. of
    Oregon, Mus. Nat. Hist., Bull.14, 67p.
j--1968b. A report to the National Park Service on the signifi-
    cance, history of investigation, and salient paleontological
    features of the upper John Day Basin, Wheeler and Grant
    Counties, Oregon (unpubl.), 27p.
k--1970. Pliocene mammals of southeast Oregon and adjacent
    Idaho. Univ. of Oregon. Mus.Nat. Hist., Bull.17, 103p.
187. Shotwell, J.A., and Russell, D.E., 1963. Mammalian fauna
    of the upper Juntura Formation, the Black Butte local
    fauna. In: Shotwell, The Juntura Basin. Amer. Philos.
    Soc., Trans., v.53, p.42-69.
188. Shufeldt, R.W., 1891. On a collection of fossil birds
    from the Equus Beds of Oregon. Amer. Naturalist,
    v.25, p.259-262.
a--1912. Prehistoric birds of Oregon. Overland Monthly, v.60,
    p.536-642.
b--1913. Review of the fossil fauna of the desert region of
    Oregon, with a description of additional material collect-
    ed there. Amer. Mus. Nat. Hist., Bull.25, p.259-262.
c--1915. Fossil birds in the Marsh Collection of Yale University.
    Conn. Acad. Arts Sci., Trans., v.19, p.1-110.
189. Silberling, N.J., and Tozer, E.T., 1968. Biostratigraphic
    classification of the marine Triassic in North
    America. Geol. Soc. Amer., Special Paper 110, 63p.
190. Sinclair, W.J., 1901. Discovery of a new fossil tapir in
    Oregon. Jour. Geol., v.9, p.702-707.
a--1903. *Mylagaulodon*, a new rodent from the upper John Day
    of Oregon. Amer. Jour. Sci., ser.4, v.15, p.143-144.
b--1905. New or imperfectly known rodents and ungulates from
    the John Day series. Calif. Univ., Publ. Geol. Sci.,
    v.4, no.6, p.125-143.
c--1906. Some edentate-like remains from the Mascall beds of
    Oregon. Calif. Univ., Publ. Geol. Sci., v.5, no.2,
    p.65-66.
191. Smith, H., 1932. The fossil flora of Rockville, Oregon.
    Univ. of Oregon, Masters, p.1-44.
192. Smith, J.P., 1912. Occurrence of coral reefs in the
    Triassic of North America. Amer. Jour. Sci., 4th ser.,
    v.33, p.92-96.
193. Smith, W.D.P., and Allen, J.E., 1941. Geology and
    physiology of the northern Wallowa Mountains, Oregon.
    Oregon Dept. Geol. and Min. Indus., Bull.12, 64p.

194. Snavely, P.D., and Baldwin, E.M., 1948. Siletz River
     volcanic series, northwestern Oregon. Amer. Assoc.
     Petroleum Geol. Bull., v.32, p.806-812.
195. Snavely, P.D., and Wagner, H.C., 1963. Miocene geologic
     history of western Oregon and Washington. Washington
     (State) Div. of Mines and Geol., Rept. of Inves., no.22,
     25p.
196. Snavely, P.D., Rau, W.W., and Wagner, H.C., 1964. Miocene
     stratigraphy of the Yaquina Bay area, Newport, Oregon.
     Ore Bin, v.26, no.8, p.133-151.
197. Snavely, P.D., MacLeod, N.S., and Rau, W.W., 1969.
     Geology of the Newport area, Oregon. Ore Bin, v.31,
     no.2, p.25-47.
198. Snavely, P.D., MacLeod, N.S., Rau, W.W., Addicott, W.O.,
     and Pearl, J.E., 1975. Alsea Formation; an Oligocene
     marine sedimentary sequence in the Oregon Coast Range.
     U.S. Geol. Surv., Bull.1395-F, 21p.
199. Sorauf, J.E., 1972. Middle Devonian coral faunas (rugose)
     from Washington and Oregon. Jour. Paleo., v.46, no.3,
     p.426-639.
200. Staples, L.W., 1950. Cubic pseudomorphs of quartz after
     halite in petrified wood. Amer. Jour. Sci., v.248,
     no.1, p.124-136.
201. Steere, M.L., 1954. Fossil localities of Lincoln County
     beaches, Oregon. Ore Bin, v.16, no.4, p.21-26.
a--1955. Fossil localities in the Coos Bay area, Oregon.
     Ore Bin, v.17, no.6, p.39-43.
b--1957. Fossil localities of the Sunset Highway area, Oregon.
     Ore Bin, v.19, no.6, p.51-59.
c--1958. Fossil localities of the Eugene area, Oregon. Ore
     Bin, v.20, no.6, p.51-58.
d--1959. Fossil localities of the Salem-Dallas area. Ore
     Bin, v.21, no.6, p.51-59.
202. Stearns, R.E.C., 1902. Fossil shells of the John Day
     region. Science, n.s., v.15, p.153-154.
a--1906. Fossil mollusca from the John Day and Mascall beds
     of Oregon. Calif. Univ., Publ. Geol. Sci., v.5, no.3,
     p.67-70.
203. Stirton, R.A., 1944. A rhinoceros tooth from the Clarno
     Eocene of Oregon. Jour. Paleo., v.18, no.3, p.65-67.
204. Stirton, R.A., and Rensberger, J.M., 1964. Occurrence of
     the insectivore genus *Micropternodus* in the John Day
     Formation of central Oregon. Southern Calif. Acad.
     Sci., Bull., v.63, pt.2, p.57-80.
205. Stock, C., 1925. Cenozoic gravigrade edentates of western
     North America with special reference to the Pleistocene
     Megalonychinae and Mylodontidae of Rancho LaBrea.
     Carnegie Inst., Wash., Publ.331, 206p.
a--1930. Carnivora new to the Mascall Miocene fauna of eastern
     Oregon. Carnegie Inst., Contrib. to Paleo., Publ.404,
     p.43-48.
b--1946. Oregon's wonderland of the past, the John Day.
     Science Monthly, v.63, p.57-65.

206. Stock, C., and Furlong, E.L., 1922. A marsupial from the John Day Oligocene of Logan Butte, eastern Oregon. Calif. Univ., Dept. Geol. Sci., Bull., v.13, no.8, p.311-317.

207. Sullivan, R.B., 1970. U.S. Geological Survey radiocarbon dates XI, Cape Fisheries, Oregon. Radiocarbon, v.12, no.1, p.325.

208. Taggart, R.E., and Cross, A.T., 1974. History of vegetation and paleoecology of upper Miocene Sucker Creek beds of eastern Oregon. Birbal Sahni Inst. Paleobot., Spec. Publ., no.3, p.125-132.

209. Taylor, D.W., 1960. Distribution of the freshwater clam *Pisidium ultramonanum*; a zoogeographic inquiry. Amer. Jour. Sci., Bradley vol., v.258-A, p.325-334.

a--1963. Mollusks of the Black Butte local fauna. Amer. Philos. Soc., Trans., v.53, pt.1, p.35-40.

210. Taylor, E.M., 1960. Geology of the Clarno basin, Mitchell quadrangle, Oregon. Oregon State Univ., Masters, 173p.

211. Terry, J.S., 1968. *Mediargo*, a new Tertiary genus in the family Cymatiidae. Veliger, v.11, no.1, p.42-44.

212. Thoms, R.E., 1965. Biostratigraphy of the Umpqua Formation, southwest Oregon. Univ. Calif., PhD., 215p.

213. Thoms, R.E., and Smith, H.C., 1973. Fossil bighorn sheep from Lake County, Oregon. Ore Bin, v.35, no.8, p.125-134.

214. Thorpe, M.R., 1921a. John Day Eporeodons, with description of new genera and species. Amer. Jour. Sci., ser.5, v.2, p.93-111.

a--1921b. John Day Promerycochoeri with description of five new species and one new subgenus. Amer. Jour. Sci., ser.5, v.1, p.215-244.

b--1921c. Two new forms of Agriochoerus. Amer. Jour. Sci., ser.5, v.2, p.111-119.

c--1921d. Two new fossil carnivora. Amer. Jour. Sci., ser.5, v.1, p.477-483.

d--1922a. *Araeocyon*, a probable old world migrant. Amer. Jour. Sci., ser.5, v.3, p.371-377.

e--1922b. Oregon Tertiary Canidae with descriptions of new forms. Amer. Jour. Sci., ser.5, v.3, p.162-176.

f--1925. A new species of extinct peccary from Oregon. Amer. Jour. Sci., ser.5, v.7, p.393-397.

215. Tidwell, W.D., 1975. Common fossil plants of western North America. Brigham Young University Press, Provo, Utah, 197p.

216. Trimble, D.E., 1963. Geology of Portland, Oregon, and adjacent areas. U.S. Geol. Survey, Bull. 1119, 119p.

217. Turner, F.E., 1938. Stratigraphy and mollusca of the Eocene of western Oregon. Geol. Soc. Amer., Special Paper 10, 130p.

218. Uyeno, T., and Miller, R.R., 1963. Summary of late Cenozoic freshwater fish records for North America. Univ. Mich. Mus. Zool., Occasional Papers, no.631, 34p.

219. Vallier, T.L., and Brooks, H.C., 1970. Geology and copper deposits of the Homestead area, Oregon and Idaho. Ore Bin, v.32, no.3, p.37-57.

220. Vanderhof, V.L., 1937. A study of the Miocene sirenian *Desmostylus*. Univ. Calif., Publ. Geol. Sci., v.42, p.169-164.

221. Vanderhof, V.L., and Gregory, J.T., 1940. A review of the genus *Aleurodon*. Univ. Calif., Publ. Geol. Sci., v.25, no.3, p.143-164.

222. Van Frank, R., 1955. *Palaeotaricha oligocenica*, new genus and species; an Oligocene salamander from Oregon. Brevoria, no.45, p.1-12.

223. Vokes, H.E., Norbisrath, H., and Snavely, P.D., 1949. Geology of the Newport-Walport area, Lincoln County, Oregon. U.S. Geol. Surv., Oil and Gas Inv. Map, OM 88.

224. Vokes, H.E., Snavely, P.D., and Myers, D.A., 1951. Geology of the southern and southwestern border areas, Willamette Valley, Oregon. U.S. Geol. Surv., Oil and Gas Inv. Map, OM 110.

225. Wagner, Norman S., Brooks, H.C., and Imlay, R.W., 1963. Marine Jurassic exposures in Juniper Mountain area of eastern Oregon. Amer. Assoc. Petroleum Geol., Bull., v.47, no.4, p.687-701.

226. Wallace, R.E., 1946. A Miocene mammalian fauna from Beatty (Beatys) Buttes, Oregon. Carnegie Inst., Wash., Publ.551, p.113-134.

227. Ward, P.D., and Westerman, G.E.G., 1977. First occurrence, systematics, and functional morphology of *Nipponites* (Cretaceous Lytoceratina) from the Americas. Jour. Paleo., v.51, no.2, p.267-273.

228. Warren, W.C., Norbisrath, H., and Grivetti, R.M., 1945. Geology of northwestern Oregon, west of Willamette River and north of latitude 45° 15'. U.S. Geol. Surv., Oil and Gas Inv. Map, OM 42.

229. Warren, W.C., and Norbisrath, H., 1946. Stratigraphy of upper Nehalem River Basin, northwestern Oregon. Amer. Assoc. Petroleum Geol., Bull., v.30, no.2, p.213-237.

230. Washburne, C.W., 1914. Reconnaissance of the geology and oil prospects of northwestern Oregon. U.S. Geol. Surv., Bull.590, 111p.

231. Waterhouse, J.B., 1968. New species of *Megousia* Muir-Wood and Cooper and allied new genus from the Permian of Australia and North America. Jour. Paleo., v.42, no.5 p.1171-1185.

232. Weaver, C.E., 1942. Paleontology of the marine Tertiary formations of Oregon and Washington. Univ. Wash., Publ. in Geology, v.5, pt.I, II, III, 790p.

233. Welton, B.J., 1972. Fossil sharks in Oregon. Ore Bin, v.34, no.10, p.161-170.

234. Wetmore, A., 1940. A checklist of the fossil birds of North America. Smithsonian Misc. Coll., v.99, no.4, p.1-81.

235. White, C.A., 1885. On invertebrate fossils from the Pacific Coast. U.S. Geol. Surv., Bull.51, p.28-32.

236. Wilkes, C., 1845-1874. United States Exploring Expedition, during the years 1838, 1839, 1840, 1841, 1842, under the command of Charles Wilkes. Lea and Blanchard Publ., 13v.

237. Wilkinson, W.D., 1959. Field guidebook (Geologic trips along Oregon highways). Oregon Dept. Geol. and Min. Indus., Bull.50, 148p.

238. Wilkinson, W.D., Lowry, W.D., and Baldwin, E.M., 1946. Geology of the St. Helens quadrangle, Oregon. Oregon Dept. Geol. and Min. Indus., Bull.31, 39p.

239. Wilkinson, W.D., and Oles, K.S., 1968. Stratigraphy and paleoenvironments of Cretaceous rocks, Mitchell quadrangle, Oregon. Amer. Assoc. Petroleum Geol., Bull., v.52, no.1, p.120-161.

240. Willis, J.C., 1973. Dictionary of the flowering plants and ferns. 8th ed., Cambridge Univ. Press, Cambridge, 752p.

241. Wilson, R.W., 1937. New middle Pliocene rodent and lagomorph faunas from Oregon and California. Carnegie Inst., Wash., Publ.487, p.1-19.

a--1938. Pliocene rodents of western North America. Carnegie Inst., Wash., Publ.487, p.21-73.

242. Wolfe, J.A., 1954. The Collawash flora of the upper Clackamas River basin, Oregon. Geol. Soc. Oregon. Country, Newsletter, v.20, no.10, p.89-94.

a--1960. Early Miocene floras of northwest Oregon. Univ. Calif., PhD.

b--1962. A Miocene pollen sequence from the Cascade Range of northern Oregon. U.S. Geol. Surv., Prof. Paper 450-C, p.81-84.

c--1969. Neogene floristic and vegetational history of the Pacific Northwest. Madrona, v.20, p.83-110.

243. Wolfe, J.A., and Hopkins, D.M., 1967. Climatic changes recorded by Tertiary land floras in northwestern North America. In: Tertiary correlations and climatic changes in the Pacific; Pacific Sci. Cong. 11, Tokyo, 1966, p.67-76.

244. Woodburne, M.O., and Robinson, P.T., 1977. A new late Hemingfordian mammal fauna from the John Day Formation, Oregon, and its stratigraphic implications. Jour. Paleo., v.51, no.4, p.740-757.

245. Wortman, J.L., and Matthew, W.D., 1899. The ancestry of certain members of the Canidae, the Viverridae, and Procyonidae. Amer. Mus. Nat. Hist., Bull.12, p.109-138.

246. Yochelson, E.L., 1961. Occurrences of the Permian gastropod *Omphalotrochus* in northwestern United States. U.S. Geol. Surv., Prof. Paper 424-B, p.237-239.

247. Zullo, V.A., 1964. The echinoid genus *Salenia* in the eastern Pacific. Paleont., v.7, pt.2, p.331-349.

a--1969a. A late Pleistocene marine invertebrate fauna from Bandon, Oregon. Calif. Acad. Sci., Proc., 4th ser., v.39, no.12, p.346-361.

b--1969b. Pleistocene symbiosis, pinnotherid crabs in pelecypods from Cape Blanco, Oregon. Veliger, v.12, no.1, p.72-73.

285

NOTES

NOTES

NOTES

NOTES

NOTES

NOTES

NOTES

NOTES

NOTES